Insects and Gardens

INSECTS AND GARDENS
In Pursuit of a Garden Ecology

ERIC GRISSELL

with photographs by
CARLL GOODPASTURE

TIMBER PRESS
Portland, Oregon

Frontispiece: A fourteenspotted lady beetle, seeking its aphid supper, roams amidst a universe of forget-me-not flowers. This microcosm, depicted in photogenic vision, scarcely exists from a gardener's point of view, yet it represents the balance found in a natural world—at least in Europe. Both flower and beetle are now naturalized in North America, underscoring the gardener's potential to weave worldly elements into the tapestry we call a garden.

Published in 2001 by

Timber Press, Inc.
The Haseltine Building
133 S.W. Second Avenue, Suite 450
Portland, Oregon 97204, U.S.A.

Designed by Susan Applegate
Printed in Hong Kong

**Library of Congress
Cataloging-in-Publication Data**

Grissell, Eric.
    Insects and gardens : in pursuit of a garden ecology / by Eric Crissell ; with photographs by Carll Goodpasture.
        p. cm.
    Includes bibliographical references (p.).
    ISBN 0-88192-504-7
    Beneficial insects. 2. Garden ecology.
    I. Title.

SF517 G75 2001
577.5'54—dc21   00-052791

The balance of nature is . . . a complex, precise, and highly integrated system of relationships between living things which cannot safely be ignored any more than the law of gravity can be defied with impunity by a man perched on the edge of a cliff.

RACHEL CARSON
Silent Spring, 1962

We dedicate this book to Rachel Carson.

Contents

Acknowledgments

Although I am a research entomologist with the Systematic Entomology Laboratory, Agricultural Research Service, U.S. Department of Agriculture (USDA), no part of this book is sponsored or supported by the USDA, nor is it considered part of the mission of the USDA for its employees to write books such as this. The text was conceived and written entirely on my own time, is not based on any unpublished research conducted by me as a member of the USDA, and does not necessarily reflect the opinions of the USDA, although I would certainly hope that it does.

The topic of this book was originally conceived by Neal Maillet, executive editor, and the staff at Timber Press. I express my thanks to the press for the opportunity to write this book. I also thank the following individuals for their input into certain sections: Gabriela Chavarria, National Fish and Wildlife Foundation, Washington, D.C., for reviewing the section dealing with insect pollinators; Robert Denno, professor of entomology, University of Maryland, College Park, for reviewing the entire section on insect ecology and making many useful suggestions for improving it; Natalia Vandenberg, Systematic Entomology Laboratory (SEL), Agricultural Research Service, USDA, for introducing us to the world of singing insects; Alma Solis, SEL, for information on predatory caterpillars; Gary Miller, SEL, for information about insects in history; Dug Miller and Tom Henry,

SEL, for help with the confounding true bugs; Ronald Ochoa, SEL, for information concerning mites; and George Venable, scientific illustrator, Department of Entomology, Smithsonian Institution, for help with the digital versions of some illustrations used in this book. I also thank Lisa DiDonato for her thoroughly expert copyediting.

During the writing process, I have had many opportunities to discuss its contents with Carll Goodpasture, whose photographs grace these pages. He has read the manuscript in various incarnations and has offered suggestions for improvement. In that sense, Carll is more than a photographer or an artist, he is a collaborator. I appreciate his generosity in supporting the text with excellent photographs and literary criticisms. Gro Heining, Carll's partner in artistic adventures, has offered much help in the book's production, and I thank her for her patience—both in dealing with me and, more often, with Carll.

Introduction

Since the dawn of gardening—before an age, some might say, when the fruit of the tree of knowledge needed chemicals to protect it— there have been many kinds of gardens and many types of gardeners. A quick look at contemporary bookshops overwhelms the senses with information. A cadre of books and garden writers attests to the notion that gardening is no longer simply a matter of doing right or wrong, it is one of unmitigated convolution. For example, should you be an organic gardener, a chemically dependent one (what is fondly referred to as a "nozzlehead"), or any one of a hundred finer shades in between? Should you be a cottage gardener, a Victorian gardener, an herb gardener, a plain dirt gardener, a natural gardener, a container gardener, a colorist, or an enlightened combination of all sorts of specialist dogmas? Should you garden an hour a day, twenty minutes a day, or even one minute a day, as some authors would tempt us to believe? Do we push the envelope and garden without a garden or even—the ultimate sin—garden without work?

I have read it all, heard it all, tried it all. It is my contention that if you wish to succeed in the gardening life, there is only one true path to salvation—be a *realistic* gardener. And that is where most of us, including myself, frequently drift off the pathway and fall headfirst into the bramble patch. We lose sight of what a garden is, what it is not, and where it wants to go.

What a garden *is*, by definition, is a humanly contrived, artificial construct of plants from all over the world that is usually assembled with little regard for ecological or biological principles. At the same time, this assemblage of plants exists in a world whose entire life force, whose every purpose of being, is driven by universal principles of environmental or ecological balance. In simple scientific lingo, a garden is essentially a nonfunctional ecosystem. If you don't like scientific terms, then a garden might be thought of as a struggle between a piece of land trying to restore itself to a natural balance and a gardener who hasn't a clue what that means. Thus begins the eternal battle between the gardener, the garden, and the forces of nature.

What a garden is *not*, by any means, is natural. Certainly, there will be those who disagree with me, some violently, but they will be wrong, of course. I am both a professional naturalist and an amateur gardener, and I ought to know the difference. There is currently a greatly self-conscious controversy awash in the concept of the natural garden—a movement that has spawned epithets and counter-epithets that occasionally make me embarrassed to call myself a gardener. (Consult Janet Marinelli's *Stalking the Wild Amaranth* for an enlightened look at some of the gory details.) The notion of a natural garden, although noble and one with which I empathize, is an ideal doomed, if nothing else, to semantic failure. In truth, a "natural garden" is simply an oxymoron.

Nature or natural, taken in the context of the biological realm—what I like to think of as the Real World—is what transpires when humankind does nothing at all, when we let the world's majestic environmental forces act at will. Natural is the absence of manipulation, of tinkering. This concept goes entirely against the tenets of humankind, which has a deep historical, hysterical, if not just plain mystical belief that it is in control of everything. With a little tinkering, everything can be made better by humans.

Foremost among control zealots is the average gardener. Can you imagine a gardener allowing the environment to act at will? The garden being left to its own care? I certainly cannot. I know of no gardener who can sit for a moment and do nothing, much less do nothing at all for the rest of his or her life, which is what a truly natural garden would entail. A typical gardener is always messing about, scarcely ever content to let nature act at all. Scurrying here, pruning there, staking, weeding, spraying, watering, fertilizing, adding plants, removing plants, altering the soil, fencing things in or fencing them out, cutting the lawn, encouraging birds, discouraging deer. The gardener's life has historically been the antithesis of nonintervention, of anything hinting at natural.

The truth is that our understanding of gardening is divorced from reality, whether it be the simplest biological fact or the most complex ecological theory. We gardeners do not apply the principles that govern biology, ecology, the environment, Mother Earth, Gaia, or whatever term you wish to call the world that has not yet been invaded and manipulated by humans. We gardeners do not take the easy, natural path if we can feasibly make everything as impossibly difficult as is humanly imaginable. Not only are our gardens dysfunctional ecosystems, but in urban and suburban regions we have largely disrupted all the natural areas around our gardens, our homes, our offices, and our cities. Rural farming areas are scarcely different, really, because a house and garden in the middle of a farm are, ecologically speaking, no different than a house surrounded by a parking lot, at least in terms of functioning in a relatively natural way.

To put the situation into perspective, few people live within the confines of an undisturbed meadow, virgin forest, or pristine lakeshore. Even if we did, the first thing we would want to do is to change something—anything—to make our surroundings somehow better. It is simply human nature to meddle. If we accept this view of ourselves, that is, if we can be realistic gardeners, then we will have established a common point from which to progress to the next level of ecological reason: *if we work with the laws of nature, we have a much better chance of developing a garden that functions as a balanced, naturalistic system should.*

I believe that the more naturalistic a garden becomes, the more enjoyable it is for the gardener; the more enjoyable the garden becomes, the easier it is for the gardener to manage; and the easier the garden becomes to manage, the more likely the gardener is to encourage it to become more naturalistic. This is a virtuous cycle that might take us from our normally high levels of mechanical, manual, and chemical input into higher levels of delight and respect for the natural order of things.

To be a gardener, you must possess an optimistic imagination combined with a sense of adaptive reality. Although we may invoke the most lofty and artistic ideas, the gardener must ultimately face the garden on *its* terms or face the alternative of constant vigil or eventual ruin. A garden of the mind, although a thing of beauty and a joy internal, comes to naught if it cannot be rooted firmly in the ground. And that is what this book is all about. Rooting your garden firmly in the groundwork that nature has established for its creatures, whether plant or animal, whether the size of an elephant or that of a virus.

The more naturalistic thinking among you may have noticed, by now, that I have said virtually nothing about insects, the purported subject of this book. Convincing readers, in general, and gardeners, in particular, that insects are worth knowing about is a difficult job, and I've been putting it off until I built up the courage to do so. There is an inherent difficulty in defending Japanese beetles, yellow jackets, mosquitoes, mealybugs, hornworms, and the few other insects that people *do* know about—usually in a negative context—while trying to explain the role of the myriad other insects they do not know about. But gardeners need to know a little about *all* the insects in their gardens if they are to understand how a garden functions properly, naturalistically, because insects, like the soil, are an essential part of a garden's structure.

To insist that insects—good, bad, or indifferent—should be banished from the garden is to begin a skirmish that leads to incessant warfare, a warfare that is unwise, unwinnable, and virtually unnecessary. According to Environmental Protection Agency statistics for 1993 alone, 1.1 billion pounds of pesticides were used by agriculture (75 percent), industry (18 percent), and households (7 percent) to rid the United States of pests. Is it possible to consider this an environmentally good approach—a healthy approach—to solving our worldly problems? Even if we win the battle in our gardens, we ultimately lose the war for the environment upon which we must all depend. The gardener's ultimate battle—our Sisyphus—should be to make the artificial become as natural as possible. The more we impose our wishes against the processes of nature, the less chance we have of living within the bounds of nature. And never forget that nature has had billions of years to create its adapt-or-die attitude; nature has a 24-hour-a-day, 365-day-a-year advantage on us mere gardeners.

So what advantage do we have, those of us who till the soil? We gardeners might respond that we have intelligence. But there are those who would counter that because we *are* gardeners, we obviously have abandoned this advantage. Whether we are intelligent or not, my belief is that the only way we will overcome the imbalance of nature in our gardens is through our own cursed orneriness and our time-honored ability to withstand abuse. (Even those who are not gardeners will concede our tenacity.) To fight the good fight, we gardeners are going to have to adapt to nature, not the other way around. In fact, we won't even have to fight—adapting is simply the natural thing to do.

If we are to set things right in the garden, to restore some semblance of ecological balance, then we had best be realistic gardeners and make an honest appraisal of how the garden works, who its players are, what they are doing, and why they are doing it. Understanding the difference between what takes place in nature, what takes place in the garden, and how to integrate the two realms is the primary goal of this book, and insects are its central players. With some basic understanding of insects and the roles they play, we gardeners will be better able to assess and respond to situations that arise in our midst, or better yet, simply learn to ignore them. If the timeless advice is true that "ignoring a problem will make it go away," then the hallmark of a great gardener may well be knowing how and when to let nature correct the gardener's own mistakes.

PART I

A swallowtail butterfly is readily admired by almost any gardener. The butterfly is one of the few insects we unconditionally accept based on its totally benign nature and ethereal beauty. This butterfly is also a lesson in adult morphological structure by displaying the three main body regions of an insect: the head, with its two antennae above and strawlike mouthparts below; the thorax, with three pairs of legs and two pairs of wings; and the abdomen.

Lives of the Insects

> Like it or not, insects are part of where we have come from,
> what we are now, and what we will be. It seems to me that's
> a pretty good reason for getting better acquainted with them.
>
> MAY BERENBAUM
> *Bugs in the System*, 1995

We gardeners look upon our gardens as the one place in the world where we might exercise some control over something—anything, really. We cannot battle our bosses; the commute to and from work is a terror; the auto mechanic is out to cheat us; the doctor wants to stick things where things ought not to be stuck; and, on a good day, we owe more money than we will ever make. Even the highly sought-after home—our life-assuring castle—is as likely to be an endless money pit as a joy forever. At best, we are left with a lawn and garden over which, we blithely convince ourselves, we have eminent domain. My fellow gardeners, let me assure you that if you cannot keep the kitchen sink from backing up or the paint from peeling, the roof from leaking or the washing machine from breaking down, you have no hope whatsoever of keeping the forces of nature away from your garden.

In the first part of this book, I will explain some simple facts about the bugs and the bees so that you, the gardener, may better appreciate why these apparently irritating life-forms are as much the backbone of your garden as is its soil. Perhaps by understanding what insects are and what they do, we will appreciate how they function both to please us and to displease us. Insects are merely insects, after all, they have no worldly pretensions other than survival. It is how we gardeners interpret their lives that will determine our reactions to them.

In chapter 1, I discuss much of the confusion surrounding insects and other creatures not even closely related to insects. Spiders are neither bugs nor insects, for example, and the reasons for this are discussed at a basic, need-to-know level. Chapter 2 is an overview of insects, their diversity, and the orders most likely to be found in the garden by the average gardener. Because knowledge is power, at least in the world of epigrams, I also present information about insects that live in the garden but will never be seen by even the best of gardeners. Some of this information is presented not so much to inform, but to impress the gardener with the knowledge that many unseen interactions transpire in the garden without any help from its master. The notion that much happens in the garden of which we are unaware, may give us a simple insight into how little the gardener actually knows her own plot of land. In chapter 3, I explore the growth and development, sex lives, and social interactions of insects to examine why insects appear to be so successful in their struggle to take over the world—at least in the opinion of most gardeners. Finally, in chapter 4, I examine the basic biological approaches that insects have evolved and how insects fit into the environment, in general. In part II, we will discover how insects fit into the environment of our gardens, in particular.

1

What Is (and Is Not) an Insect?

All animals, including humans, must distinguish between certain elements critical to their personal survival: between food and non-food, between friend and foe, between mate and nonmate, between shelter and a viper's pit. Not to do so could certainly cause great inconvenience, as, for example, if a cow mistook a hornet's nest for supper, a rabbit mistakenly befriended a wolf, or a mouse attempted to mate with an elephant. No less than the survival of the species relies on an individual's ability to recognize the essence of its natural history—to know what it is, where it is, and how to fit into its environment.

The ability to sort an environment into its critical elements comes naturally to all animals, even if they do not consciously realize they do it. Predators recognize prey, for example, and plant feeders recognize their host plants. We humans have adapted our ability to recognize things to a fine art, particularly in the case of our fellow humans. We can, for example, instantly pick out our parents from any of six billion people on the planet, yet, to a biologist nearly all people are the same. Conversely, a biologist can immediately tell the difference between a million different kinds of insects and a spider, yet, most people can scarcely recognize ten insects, much less realize that insects and spiders are not the same things.

The art of recognition is greatly dependent on our priorities, and most humans would readily agree that recognizing our parents is more important than distinguishing insects from spiders. In the garden, of course, this is not true. Well, not usually, anyway. Our priorities change and so should our abilities to distinguish not only insects from spiders, but beetles from bugs, butterflies from bees, and dragonflies from flower flies. It may seem somewhat pedantic to say, for example, that a scorpion or a spider is not an insect, but they are not, and they should not be confused with insects. Presumably you would not call your Aunt Evelyn a snake, at least to her face, which is very much like calling a spider an insect, or even less correctly, a bug. Insects and spiders are as distantly related to each other as are mammals and reptiles. Therefore, I will be precise in my terminology, at least at a level where it matters to us gardeners.

Biologists arrange all animals and plants into hierarchies based on presumed relatedness, which is usually inferred from similarities in form and structure. The most inclusive category is a kingdom, which can be divided into phyla, classes, orders, families, genera, and finally species (or even subspecies). All insects are placed into the single class Insecta (sometimes called Hexapoda). The organisms gardeners most commonly confuse with insects are the spiders, mites, and ticks, which taken together form their own class, the Arachnida. Both groups are members of the phylum Arthropoda. See chart on pages 67–70.

If the fact that mites or ticks are not insects seems like a piffling frivolity that is inconsequential to the gardener, consider the fact that mites generally cannot be killed by insecticides, but require miticides for control. This is because the two broad groups of organisms—insects and arachnids—have different physiological, mechanical, and genetic systems and react differently to chemicals used to control them. The commercial manufacture of both insecticides *and* miticides proves that we need to know the name by which something is called, even if for no other reason than to kill it (which, by the way, we will not dwell on in this book).

The Insects

For the purist, adults of the class Insecta can be defined precisely by certain characteristics shared by all members of the class. All adult insects have six legs originating on the thorax, or midsection, of a body that is divided into three parts. There is an unwritten axiom in biology that almost no observation can be said to hold true in all cases, but for practical purposes, no other animal can be said to have three pairs of legs arising from its midsection. A second characteristic is that insects have one pair of antennae, although this is usually more esoteric than useful to the gardener's eye. A third characteristic is that adult insects are the only invertebrates (animals without backbones) that can fly, but in this case the biologist's axiom holds true, because there are a number of wingless adult insects.

The classification of Insecta into subgroups such as orders and families may be a bit more than most gardeners want to know. For now, I will say simply that there are approximately thirty orders of insects that include broad groupings such as beetles (Coleoptera); moths and butterflies (Lepidoptera); and wasps, bees, and ants (Hymenoptera). These orders, in turn, are subdivided into hundreds of families, thousands of genera, and millions of species—enough names to suggest why entomologists are often considered insane.

Because the remainder of this book treats the insects, I shall skip them momentarily and move on to those garden denizens with which they might be confused.

Noninsects in the Garden

Most gardeners know the difference between what they call "insects," or "bugs," and such other wildlife as slugs, worms, chipmunks, bats, deer, snakes, frogs, pink flamingos, and so forth. There is no need for me to comment on these noninsects.

In addition to pinpointing the distinguishing characteristics of each group of insectlike noninsects, I will tell you what they are doing in your garden in as few words as possible. (In many cases, we have only a general idea of what these creatures do, so a few words are about all that is possible.)

Millipedes and Centipedes (Classes Diplopoda and Chilopoda)

Millipedes and centipedes are more closely related to each other than to any of the groups discussed in this chapter. The most striking distinction of both is that their bodies are essentially a long tube composed of numerous, ringlike body sections. What distinguishes a millipede from a centipede is that the former has two pairs of legs per body segment, whereas the latter has only one pair. This distinction works only when you can see the legs, of course, and that poses a problem. Generally, when you disturb a centipede, it goes charging off as fast as its thirty to one hundred legs—depending on the species—will carry it. A millipede, however, tends to curl up in a ball with its sixty to two-hundred-plus legs hidden. Although the average gardener may be too busy to take a leg census, we can usually separate the two groups by the behavior just noted as well as the fact that millipedes are round like a stick, whereas centipedes are flat like a millipede someone just stepped on (which is probably the first reaction the gardener has upon turning over a rock, board, or pile of month-old compost). Millipedes and centipedes prefer to live under things, near or in humus-covered soil.

As an interesting aside into evolutionary efficiency of structure, the millipede, with twice as many legs per segment as the centipede, generally has half as many body segments. It's as if a millipede is simply a centipede that crashed headfirst into a tree and jammed every other segment with the next. It is likely that there is some other, evolutionary explanation, but fancy is often a better mnemonic than fact.

As to their roles in the garden, these usually cryptic groups have two basically different lifestyles. Centipedes are predators with well-developed, poison-injecting jaws. They feed on insects, spiders, slugs, earthworms, and other small creatures found in a well-diversified garden. Most millipedes, however, are scavengers that feed on

Millipedes are primarily scavengers that feed on decaying plant material, but some attack living plants and a few are predators.

Centipedes are predators of other small creatures and some can bite humans—painfully so.

decaying plant material. Some species are known to attack living plants and a few are predators. Millipedes are neither biters nor aggressive, but centipedes can bite humans—the bigger the centipede, the harder and more painful the bite. There is a common centipede occasionally found in our houses (the house centipede, oddly enough) that is at once the most delicate and hideous appearing creature ever conceived. It hunts down and kills small insects and spiders that linger in the home (sorry to inform you of this fact, but we all unknowingly share our living spaces with an assortment of other species). In my house, I never kill these creatures, but when I see them I do what any right-thinking person would do—faint. When I wake up they are gone.

Sowbugs and Pillbugs (Class Malacostraca)

Sowbugs and pillbugs are crustaceans, and are therefore most closely related to shrimps, lobsters, and crabs. They are the only members of this class that live out of water and in our gardens. Sowbugs and pillbugs are much like centipedes in having relatively many body segments, at least five, each with a pair of legs, but they rarely have more than a total of seven pairs. Thus, they have fewer legs than either centipedes or millipedes, but sowbugs and pillbugs also differ in having two pairs of antennae instead of a single pair, which is pretty academic stuff. It is likely that a gardener can distinguish sowbugs from centipedes based more on size and color than on exact characters. Sowbugs and pillbugs are generally short and gray or pearlescent, whereas centipedes are long and reddish orange and millipedes are long and black or darkly colored.

Sowbugs and pillbugs have always been thought to feed on decaying vegetable matter and thus to be fairly innocuous. Occasionally, however, they feed indiscriminately on a diversity of seedlings. Research has muddied the picture a bit with the recent discovery that pillbugs also act as predators by feeding on stinkbug eggs. Stinkbugs, themselves, are serious plant feeders. How the gardener views pillbugs now must be tempered with some caution. Perhaps they are generally more predatory than we give them credit. Although sowbugs and pillbugs can be exceedingly abundant, they must have a moist or wet substrate on which to live because they breathe through gills, just like their cousins the shrimps and crabs. Because these creatures are nocturnal, you rarely see either unless you pick up a flowerpot or turn over a board lying in the garden.

Distinguishing between sowbugs and pillbugs may not be worth the trouble because they are ecologically equivalent garden inhabitants. Technically speaking, a sowbug is the one that does not roll into a tight little ball when it is disturbed, but the pillbug does. Also,

Sowbugs (shown here) and pillbugs are virtually indistinguishable. Both are scavengers, but they may attack living plants, especially in the seedling stage. Research has shown that they also feed on the eggs of plant-feeding insects.

sowbugs have a pair of terminal, tail-like appendages that are lacking in pillbugs.

Spiders, Scorpions, Daddy-Long-Legs, Ticks, and Mites (Class Arachnida)

It usually comes as something of a surprise when we're told that these common garden critters are all classified in the same broad grouping. What they all have in common is eight legs (four pairs) originating on what appears to be the front half of a body that is divided into two sections. Relative to insects, arachnids have the head fused to the thorax (that is, one less body part), which contains an extra pair of legs. They also have no antennae.

Most gardeners are familiar with spiders based on their appearance and their most obvious work, the web, although not all spiders make webs. All spiders are predatory on small organisms, and insects form a large part of their diet. Some spiders are known to feed on birds, frogs, and small lizards, but we need not delve into this rather scary notion. From the gardener's viewpoint, spiders are our friends. Because this is not a book about spiders, I will not discuss the ostensible problems of black widows, brown recluses, or the webbed trapping of butterflies, honey bees, or other beneficial insects. The whole concept of good and bad in the garden must wait until later in our adventure, when I shall dispel the myth almost entirely.

Scorpions are predators that feed on insects and spiders. They are not as commonly encountered as spiders. Part of the reason is because they are basically nocturnal creatures, preferring to remain under rocks and logs during the day and to hunt at night. Another reason is that there are relatively few species of scorpions (perhaps fifty in the United States) compared to spiders (perhaps 75,000 worldwide). In spite of their reputation, only a few species of scorpions have the ability to actually kill a human with their sting. These species live in Arizona, a fact that I am certain does not crop up much in those fanciful retirement brochures.

Daddy-long-legs, or harvestmen, are one of those improbable

creatures of the animal world: a body suspended in midair from eight fragile bits of scaffolding. For nearly fifty years, I have looked at daddy-long-legs—overlooked them in truth—without a single notion as to what they do in the garden. Although no one knows precisely, they appear to be scavengers, predators, or both. According to the elegant musings of Sue Hubbell (1993), her pet daddy-long-legs lived on a diet of "persimmons, cornmeal, bacon fat, and an occa-

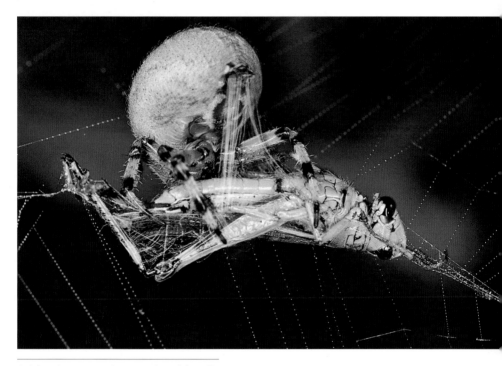

Spiders are everywhere and could easily form the basis of an entire book on interactions in the garden. Contrary to popular belief, spiders are not insects. For that reason they are not treated extensively in this book, but they are among the most voracious predators in the garden. Shown here is a garden spider feeding on a grasshopper.

The biology of daddy-long-legs, or harvestmen, is virtually unknown. They are thought to be scavengers, but they may be predators. The next time your local PBS station shows its thousandth special on the tropical rain forests, call them and ask what daddy-long-legs are doing in their own backyard. It will give them something to think about.

sional dead fly," which is not much different than my own midday meal. If you think that all the great challenges to biology exist either in tropical rain forests or in the African veldt, think about the daddy-long-legs you've seen bouncing along your pathways. Odd, isn't it, how we humans know more about elephants in far off places than we know about life in our own gardens?

We tend to know a lot more about ticks than daddy-long-legs, but this is merely a case of self-preservation. Ticks are blood-sucking parasites that feed on mammals, birds, and reptiles. Unfortunately for us, gardeners are mammals. Ticks are rarely a concern for gardeners unless we happen to live in rural or semirural areas. These creatures, the ticks not the gardeners, appear to be short-legged spiders upon which some unkind person has dropped a brick. That is, they are compressed like a wafer until they fill up with blood, at which point they expand, drop off, and digest what used to be inside of you. And that's the good thing about ticks. The bad thing is that they may carry some sixty or more pathogens that cause such discomforts as Lyme disease, Rocky Mountain spotted fever, relapsing fevers, tularemia, and tick paralysis. As garden critters go, ticks are not particularly appealing, and even though I will eventually say positive things about mosquitoes, wasps, bees, and even aphids, I cannot think of much that will comfort us when it comes to ticks.

Mites are one of the minuscule concerns that gardeners face from time to time, without even knowing it. There are many kinds of mites, including plant feeders, gall formers, predators, and parasites, all of which live somewhere at some time in our gardens. Those who live in the Midwest and South are unhappily familiar with harvest mites, or chiggers, whose immature forms attack humans by sticking their mouthparts into our skin and secreting a fluid that partially digests the tissues. The usual reaction, to which I can freely attest, is unbearable itchiness that lasts for days.

Although chiggers are nuisances, gardeners are more apt to run into a problem of plant loss than skin itch. Suddenly a plant (especially in the house) becomes festooned with webs and begins to look spotted and bedraggled. Then it dies. But because mites are the size

31

of a pinhead (or sometimes much smaller), their presence most often goes unnoticed and the cause of death is unsuspected. This is basically a good thing, because mites surround us. But because we don't readily see mites, we seldom have cause to worry about them. As you will soon discover, one of the basic tenets of this book is that, more often than not, the gardener's best defense is a basic laissez-faire approach of accepting insects and noninsects in the garden.

And well it should be, for I will now tell you something that will stick with you the rest of your life, literally. All humans have a species of mite that lives in the follicles of the eyebrows on our faces.

Mites are tiny. Chances are the average gardener has never seen one. Here, a daddy-long-legs is carting mites (the red blobs) around on its spindly legs. The mites are probably feeding as true parasites, sucking vital fluids from their host.

I'll bet you didn't want to hear that bit of news! But wait, there's more: another species of mite lives off the glands that branch from the follicles of these same eyebrows. That's right, we have two species of mites living on our faces, and there is nothing you or I can do about it. So, when I warn that it is better to be a laissez-faire gardener and to turn the other cheek (or eyebrow), you had better listen to me or else I'll have to tell you something that might really get your attention. I don't want to have to be forced to do that.

2

Orders in the Garden

Humans use the term *animal* freely, often without much degree of precision. The word *animal,* for most people anyway, probably is associated with concepts of wolves and dogs, leopards and cats, or warthogs and pigs. Birds, of course, are not animals they are birds; fish are fish. Snakes, bats, and other less appreciated life-forms become lower than animals to many—they become vermin. Somewhere at the bottom of the heap lie the insects. Even butterflies, the darlings of the gardening set, are not often thought of as animals, that is, in the same breath as something like a chipmunk.

When I was taught biology in the 1960s, there were basically two categories of life-forms: plants and animals. Humans fell into the category of animals, and we were jumbled up with all the millions of life-forms whose cell walls did not contain cellulose, as did plants'. In present times, distinctions have become more precise and we now recognize five different kingdoms of basic life-forms: Monera (bacteria and viruses, for example), Protista (algae, diatoms, and slime molds), Fungi, Plantae, and Animalia. In spite of these divisions, humans still fall into the category of Animalia with all the rest of the animals from slugs to whales—and insects too.

As far as animals go, biologists presently (for the time being, at least) consider insects to be the most biologically diverse, most numerically abundant, and most speciose of all life-forms on the planet. I use the term *presently* because relatively poorly studied groups of tiny creatures such as nematodes or bacteria could be found to be superabundant. It's hard to know. For now, however, we award the prize to insects in all their glorious abundance. They appear to deserve it, as should soon become apparent.

Numbers

Insects, at an estimated weight of 27 billion tons, outweigh the human population by about six times. In terms of biological mass, or biomass, insects are by far the dominant animal life-form on Earth. I'm not certain how people find the time to make these measurements, but I guess they might nearly be true. An estimated total of 1 quintillion individuals, at any point in time, suggest that insects outnumber humans by a lot more than they outweigh us, but we sort of suspected this already. In terms of the Earth's biological diversity, or biodiversity (by some authorities the same as species diversity), insects presently account for 80 percent of the world's animal species. Some 800,000 to 1 million species of insects have been described, and estimates put the likely figure of those species remaining to be discovered, if they are not exterminated first, between 2 million and 30 million. Clearly, there are a lot of insects in the world. Whenever we gardeners feel terribly put upon, we might consider ourselves lucky that we have so few insects with which to contend.

Order in the Garden

Insects are subdivided into groups of related sorts, called orders, based on selected morphological features. These features, or characters (as they are called in the study of relationships), are the shared traits that all members have derived from a common ancestor through the process of evolution. Thus, theoretically at least, closely related groups share one or more characters that have evolved over eons of time. Systematists, of which I am one, study insects as well as other animals and plants in an attempt to classify organisms based on their accumulation of these shared, derived characters.

This is neither easy nor always correct, but in the case of insects, entomologists have developed a fairly clear idea of what the main groups are and how to define them.

Many gardeners, in a nonscientific way, do much the same. When we see a butterfly, for example, we know that it is a butterfly. When we see a moth, we might not know exactly why it is a moth, but we know that it seems pretty much like a butterfly and not very much like a beetle. Moths and butterflies are closely related and form a single order of insects, Lepidoptera, whereas the beetles form another order, Coleoptera. As gardeners, then, we already have at least a semblance of knowledge when it comes to the different orders of insects. In this chapter we will hone that knowledge just a bit, adding to it the orders we will likely find in the garden—after which, I will throw in all the rest just for good measure. I present the orders of insects in the order of likelihood they will be seen in the garden by the average gardener.

It would be impossible to discuss all insects in detail or even superficially, for that matter. The subject is immense and would take thousands of books and scientific articles to do so. Because of this, entomologists are highly specialized. The best I can hope to do in this chapter is to present an overview of insect orders and how they relate to gardeners. In these discussions, I present some distinguishing features, how particular insects develop (see chapter 3 for a formal explanation of insect developmental stages), their numbers, what they eat, and where they might be found in the garden. This chapter highlights insects as organisms in their own right, whereas chapters 6 and 7 give examples of how insects interact with plants and with each other in our gardens.

The identification of insects below the category of order is not especially easy without learning a great deal about what some might call "fiddly bits." After the ordinal level comes the family level, of which there are nearly 600 in North America alone. A reasonably good beginner's guide to insect families is *A Field Guide to the Insects of America North of Mexico* by Borror and White (1970). It is a start, at least, but the study of insects is a vast subject that is fraught with an endless array of details—details that, I might add, are endlessly fascinating for those who wish to pursue the subject. We gardeners will be able to barely graze the subject's surface as we wander through our gardens looking at the various insects along the way.

Commonly Found Orders in the Garden

Lepidoptera: Moths and Butterflies

There is likely not a gardener alive who does not know what butterflies are. Butterflies, as we know, fly about the garden during its sunlit hours, embellishing an otherwise splendid picture with added color and charm. Moths, in contrast, are most commonly night fliers and tend to be drab when compared to their fairer sisters. Also, we most often find butterflies resting with their wings held vertically over their backs, whereas moths fold their wings rooflike over their backs.

Technically, the difference that distinguishes moths and butterflies from all other insects is a combination of four membranous wings (two pair), each of which is covered by thousands of small, overlapping, shinglelike scales. Scales are the things that rub off on our fingers when we try to pick up a butterfly. They may appear fine as flour, but in reality these scales are distinct structurally and provide all the color to a butterfly's wing.

LIFE STAGES: Moths and butterflies undergo a complete metamorphosis with four life stages: egg, larva, pupa, and adult.

NUMBER OF SPECIES: Worldwide there are some 140,000 to 150,000 described species.

FEEDING HABITS: There are two distinct feeding life stages. Adult moths and butterflies have specialized sucking (siphoning) mouthparts, collectively called a proboscis, which is a coiled tube used primarily for obtaining nectar from deep inside flowers. This tube is so long that it must be coiled up like a spring under the head. Adults also use the proboscis to feed on sugary substances, such as ripe

A tiny skipper (Lepidoptera) is not what most gardeners think of when the subject of butterfly gardening arises. Skippers are common (and commonly overlooked).

A colorful saturniid moth (Lepidoptera) is not a common subject of the
butterfly garden literature. Although
a moth is not a butterfly, most gardeners would be hard pressed to tell
the difference.

fruits, and to obtain moisture. Some adult moths and butterflies are known to feed on fluids that collect around the eyes of animals, and one or two even pierce the skin of animals and suck their blood (just like mosquitoes). Larval moths and butterflies are called caterpillars, with which gardeners are quite familiar. Caterpillars have jaws and use them to shred plants. These larvae are among the gardener's most common foe in the battle over plants. Caterpillars of different species have many different feeding habits. Some feed openly on leaves; others tie them or roll them up, where they feed in secret. Some species feed on roots, bore into plant stems or fruits, or make leaf mines. Others create galls or make decorative bags of plant parts in which they hang and feed on plants while being nearly completely protected. A very few larvae are predators, feeding on other insects.

GARDEN HABITS: Moth adults are most often noticed at night coming to porch or window lights. Adults are seldom observed in the garden, and even such gigantic larvae as the tomato hornworm, which are camouflaged to the hilt, are not detected without some difficulty. A few relatively well-known larvae, such as tent caterpillars and webworms, create web-tangled branches, which are conspicuous enough to arouse anyone's attention, even though the larvae themselves may not be noticed at all. Butterfly adults and larvae are generally associated with plants. Adults are most often seen in open, sunny spaces, usually flitting between the flowers upon which they feed. Their larvae, which are often camouflaged, are not so easily spotted unless they are chewing holes in big leaves, in which case it is the hole that is first seen.

Hymenoptera: Wasps, Bees, and Ants

Next to butterflies, gardeners are probably most familiar with wasps, bees, and ants based on a relatively negative (and uneducated) image. Wasps and bees are known because they sting, ants because they are everywhere (and some, like the fire or harvester ant, can sting like the dickens, too). As we shall see later in the book, gardeners are in great need of an overall attitude adjustment, especially concerning

Hymenoptera, but for now we will merely review the order and not debate its merits. There is basically no easy way to talk collectively about members of the order because they are diverse not only in name, but also in appearance, habits, and biological complexity. The term *wasp*, for example, can refer to a number of different sorts. For practical purposes, Hymenoptera are divided into the following five major categories: sawflies and wood wasps, parasitic wasps (or parasitoids), solitary wasps (or stinging wasps), bees, and ants. Collectively the only way to speak of this group in the broad sense is to call them Hymenoptera or hymenopterans.

Wasps, bees, and ants are distinguished from other insects by the presence of four membranous wings (two pair) that have no scales and are held in a horizontal plane over the abdomen when at rest. There are a few other technical differences as well, but they are of little practical use. The more astute (and technically argumentative) gardener may politely suggest that ants do not have wings at all and therefore cannot be Hymenoptera, for which I would give them a gold star for effort. Unfortunately, nothing about insects is quite as simple as it might seem. Ants have a reproductive stage that does have wings by which they disperse to colonize new areas—then they conveniently lose their wings when they go underground.

The order Hymenoptera has two rather unusual distinctions. One is that it exhibits the most diverse types of social behavior of all insects. All ants, a few bees (honey bees, bumble bees, and some tropical species), and a few wasps (yellow jackets, paper wasps, and hornets) have well-developed colonies involving reproductive and working castes, subdivided labor, and overlapping generations. The only other group of insects that lives in large colonies is the infamous termites, which we will get to momentarily. The second distinction of wasps, bees, and ants is that they have a complex method of reproduction called haplodiploidy, in which males arise from unfertilized eggs and females arise from fertilized eggs (see chapter 3, Parthenogenesis).

LIFE STAGES: All wasps, bees, and ants undergo a complete metamorphosis with four life stages: egg, larva, pupa, and adult.

Flowers and bees (Hymenoptera) were meant for each other, yet, gardeners sometimes find the association not entirely to their liking.

NUMBER OF SPECIES: Worldwide there are about 130,000 to 150,000 known species. Some experts believe this figure might eventually approach 1 million.

FEEDING HABITS: Adults feed in a limited number of ways, but their larvae have perhaps the most complex array of feeding styles among insects. Adult bees and most wasps (of all types) are nectar and pollen feeders. Many adult wasps collect insect prey to feed their young, but usually do not consume the prey themselves. Depending on the species, adult ants may be predators, seed feeders, or omnivores, feeding on almost anything they happen across. Larvae, de-

Wasps (Hymenoptera) are even more suspect than bees in the garden. This eumenid wasp hunts moth caterpillars and feeds them to her young. Here she sits atop an umbel of small flowers as she feeds on nectar.

pending on the type, may feed in any of the following ways: on plants as free-living larvae (sawflies); in plants as borers (wood wasps), gall formers (cynipid wasps), or seed feeders (seed wasps); in dead tree trunks (carpenter ants); on pollen balls provided by the adult (solitary bees); on nectar provided by the adult (social bees); on fungal spores provided by the adult (some ants); on chewed up insects provided by the adult (social wasps, some ants); or as parasites in or on other insects (parasitic wasps). In this last group, Hymenoptera excel, having more species of parasitic habit than any other group of insects.

GARDEN HABITS: In spite of the mortal (and exaggerated) fear of stings, most adult wasps, bees, and ants remain unseen to the casual observer. Ants, which nest in the soil or in rotten logs, are pretty much everywhere and can be found wandering around the garden looking for their next meal. Bees and solitary wasps are usually found at flowers sipping nectar. In late summer or early fall, when yellow jacket colonies have built up to large numbers, they are obnoxiously abundant and fearsome at our garden parties. For the most part, though, hunting wasps and parasitic wasps are invisibly scattered throughout the garden, where they are looking for hosts. The larvae of wasps, bees, and ants are extremely cryptic and remain hidden from view with few exceptions. Rarely, a cluster of pine sawfly larvae, which look exactly like moth and butterfly caterpillars, will be found, but they remain unnoticed until the gardener sees her mugo pine being skeletonized. Occasionally, when a board or rock is turned over, larvae and pupae of ants will be seen being whisked off as the frantic adults scurry them to safety. It may not be an exaggeration to say that parasitoids will be found abundantly in most gardens, where they help to maintain a balance of insect diversity, which is good for the garden, as we shall explore in part II.

Diptera: Flies

Flies are one of those things gardeners know when they see, but as with wasps, bees, and ants, most flies remain hidden from view during their lives. Gardeners will likely be familiar with mosquitoes, house flies, and perhaps syrphids (or flower flies), but the last are easily mistaken for bees because that is what they look like. In fact, one species resembles a bumble bee so closely that few would dare catch it in their hands, even though flower flies pose no threat at all.

Flies are distinct from all other insects in having one pair of membranous forewings and in having the hindwings reduced to clublike knobs called halteres. These act as gyroscopes and help the flies retain their balance as they move through the air.

Life Stages: Flies undergo a complete metamorphosis with four life stages: egg, larva, pupa, and adult. In a few rare cases, one or more of these stages is lost or appears to be lost. In one group of fungus-feeding flies, for example, pupal and adult stages can be skipped for several generations. In these flies, female larvae produce eggs within their bodies; these hatch (without fertilization) into larvae that eat their own mother. This process is known as paedogenesis. Some adult females internally retain their eggs, larvae, or even pupae until a suitable place to put them is found. In this case, no stages are really lost, they are simply hidden away inside the female.

Number of Species: There are about 120,000 described species of flies in the world.

Feeding Habits: Adult and larval flies feed in different ways on almost anything imaginable. Most adults act much like sponges, lapping and sucking up moisture and nutrition from innumerable sources ranging from nectar to feces to dead bodies. Adult mosquitoes, blackflies, stable flies, horse flies, and others suck blood from vertebrate hosts such as ourselves. Some fly adults are predators on insects, pouncing on them in midair and sucking their bodies dry. Many fly larvae (often referred to as maggots) are vegetarians that mine leaf tissue, create plant galls, and eat seeds; others eat fungi and even algae growing underwater. Several fly larvae are predators and include among their required prey such things as aphids, scales, termites, bees, ants, beetles, caterpillars, grasshoppers, millipedes, spiders, and even freshwater sponges and snails. Fly larvae are also among the most disgusting of all feeders (at least to us, anyway). They burrow deliciously through animal dung, rotting corpses (of all kinds), and dead plant tissue. Perhaps equally distressing is the fact that several groups of fly larvae feed within the bodies of living animals—in the nasal passages of sheep, for instance, or the muscles of animals (including ourselves).

GARDEN HABITS: Flies provide many vital, although thoroughly underappreciated, functions in the garden. Because their larvae eat unwanted waste such as dead bodies and fecal material, flies can be considered one of the two great orders of insect vacuum cleaners (beetles being the other). Fly larvae also help decompose dead plant material. In addition to running one of the garden's premiere janitorial services, some fly larvae are also beneficial in consuming common plant feeders such as aphids and a multitude of other insect

Flies (Diptera) are poorly thought of, in general. This furry little robber fly is not without some charm, and it eats other insects as well.

hosts. Oddly, the fly family Syrphidae, the same group that contains aphid predators, also contains several bulb flies with larvae that burrow into daffodil, amaryllis, tulip, iris, onion, and other bulbs. As with many insect groups, it is often not possible to state categorically or with certainty that they are good, bad, or indifferent with respect to the garden: some groups can fill all the niches, depending on their vast array of different species.

Coleoptera: Beetles

When someone mentions beetles, many people think of Volkswagens. These cars are nothing more than ladybird beetles with wheels. Considering the vast number of beetles, their species are surprisingly well concealed, with a few exceptions, such as the Japanese beetle and any number of ladybugs with which gardeners are well acquainted.

Beetles are characterized by having a hardened, or sclerotized, pair of forewings that cover a membranous pair of hindwings. The hardened forewings have no visible veins; when in flight, they are held upright while the body is propelled by its hindwings (somewhat the reverse of flies). No other insect group has such an arrangement of wings.

LIFE STAGES: Beetles undergo a complete metamorphosis with four life stages: egg, larva, pupa, and adult.

NUMBER OF SPECIES: The number of beetles is staggering—nearly 400,000 described species. The order contains the greatest number of species of any group of animals in the world.

FEEDING HABITS: Adult and larval beetles consume just about anything that can be thought of and some things that cannot. One beetle, for instance, is called the lead cable borer because, when it is not following its normal wood-boring habits, it bores through the sheathing that covers aerial telephone cables. No one has a clue as to why the beetle does this. With regard to normal food, beetles know no limits. In general, beetles are plant and fungus feeders, animal feeders, and scavengers and are rarely parasitic. Plant feeders eat all parts of plants from leaves to roots, stems to tree trunks, seeds to fruits. Many species feed on stored food products and clothing and carpets made of organic materials. There are cigarette beetles, carpet beetles, ship-timber beetles, sap beetles, drugstore beetles, potato beetles, cucumber beetles, and plant-boring beetles of all kinds. Adults and larvae of some species consume all sorts of dead and decaying animals and plants. In the case of animals, they eat not only the decaying flesh, but such gourmet bits as the hair and skin of

This click beetle (Coleoptera), in lift-off position, shows all its distinguishing characters. The two forewings are shell-like, and the two hindwings are membranous. We rarely see beetles in flight, but most do fly.

48

dead animals and the droppings of almost all vertebrates. There are species that survive on live animals including kangaroos, rats, beavers, and sloths, where they feed on skin secretions, hair, and various bits of skin debris. Coleoptera is one of the few orders that has many different kinds of aquatic species as both adults and larvae. Considering how diverse they are, it is somewhat odd that beetles have scarcely any parasitic forms.

GARDEN HABITS: With such a diversity of feeding habits, beetles may seem to be everywhere and in everything. This perception is not far from the truth. In the garden, many species are found under tree bark (in dead trees, if allowed to stand) and in both live and dead branches. Beetles often secret themselves under stones, boards, and fallen debris, preferring to roam the night looking for food. Because some beetles live in awful (or offal) stuff, such as animal feces and dead bodies, they perform a great service in cleaning up our gardens. Alternatively, beetles are one of the vegetable gardeners worst nightmares, having innumerable forms, many of which are specific to one or several kinds of plants (for example, potato beetle, bean beetle, cucumber beetle, asparagus beetle). We ornamental gardeners, too, must suffer the well-known Japanese beetle and rose chafer as unwelcomed guests. But perhaps the two best-known sorts of beetles make up for all the rest. One is the beloved, cheerfully cherry red ladybug, a gardener's true friend—not only beautiful to look at, but a vicious killer (of aphids) as well. The other is the lightning bug, or firefly. No one, young or old, who has ever witnessed the wonder of these insects on a warm summer's eve can say that these "bugs" are not without an uplifting and, dare I say, enlightening quality.

Hemiptera: True Bugs, Scales, Aphids, Leafhoppers, Whiteflies, and Cicadas

The saying that "All bugs are insects, but not all insects are bugs" is true. In this section we meet the true bugs and their relatives. The order is a bit tricky to understand. Sometimes the true bugs are put in one order, Heteroptera, whereas the scales, aphids, leafhoppers, whiteflies, and cicadas are placed in another, Homoptera. These latter names pop up from time to time in the gardening literature, but they are no longer used by the scientific community. (To do so requires a certain amount of unlearning, especially for us old-time entomologists.) The names, however, do serve some communicative function, so for purposes of discussion, I am going to call each group by its semiscientific, informal name: "heterops" for the true bugs, and "homops" for the scales, aphids, leafhoppers, whiteflies, and cicadas. This will be easier on the gardener and on me as well.

Heterops are distinguished by having two pairs of membranous wings with the front pair thickened at the base and both pairs overlapping in a horizontal plane. Homops have two pairs of membranous wings, but the front pair is not thickened at the base and they are held rooflike over the thorax and abdomen. Many female homops are strangely modified, perhaps none more so than the familiar scales and mealybugs that attack our plants. These forms do not have wings and, in fact, do not look much like insects at all (except for the males, which are rarely seen). It is this sort of complexity that makes tough going for the novice entomologist—and not infrequently, even the professional. Sometimes the mere appearance of an insect such as a scale is so striking for its uninsectlike facade that it is easier to learn an overall gestalt than any technical definitions. Scales and mealybugs are a fine example of this approach to learning.

LIFE STAGES: Most Hemiptera undergo a simple metamorphosis with three basic life stages: egg, nymph, and adult. In heterops, the nymphal and adult stages appear almost identical, except that the nymphs show gradually developing wing buds until they are full grown in the adult stage. Homops, however, show several different types of development. Some homops give live birth, some have alternating generations (which appear morphologically different), and some have sexes that differ in development and might not seem as if they are even in the same order.

NUMBER OF SPECIES: The order has approximately 82,000 known species, of which 32,000 are homops and 50,000 are heterops.

FEEDING HABITS: All Hemiptera have piercing-sucking mouthparts (except that adult males of some groups do not have mouth-

The order of true bugs (Hemiptera) is currently recognized as a diverse group of
insects including everything from cicadas ,
to stinkbugs, aphids, whiteflies, and scales.
What used to be called the order Heteroptera (heterops in this book) are the basic
true bugs, of which a well-camouflaged
stinkbug (shown here) is a typical example.

Within the order Hemiptera, what used to be the order Homoptera (homops in this book) include the aphids, scales, whiteflies, leafhoppers, planthoppers (shown here), and others.

parts at all) and are specialists at draining things. Most heterops are plant feeders that suck plant juices, but some are predators that pierce their chosen insect prey and pump them dry. A few bugs, like bedbugs, even suck blood from humans, bats, and birds. Homops, in contrast, are exclusively plant feeders.

GARDEN HABITS: Because the order contains many, many plant feeders, this group is among the most troublesome to gardeners. Who has not battled a miraculously expanding aphid population, for example? And if you have houseplants or a greenhouse, scales, mealybugs, and whiteflies cannot be far behind. Life for the vegetable gardener is made even worse because this group of insects, especially the homops, is the great disease spreader of the garden. What mosquitoes and disease are to humans, leafhoppers and aphids can be to plants. Viral infections with quaint-sounding names such as mosaics, yellowing, dwarfing, club leaf, striping, leaf curl, and wilts are all caused by various leafhoppers, planthoppers, aphids, and whiteflies. So, too, are a few fungal infections, such as cankers, and bacterial infections, such as blight. Although the group appears to be a crawling advertisement for sickness, there are some beneficial species as well. Predators with names like assassin bugs, pirate bugs, and ambush bugs might defray, in a small way, some of the group's more ardent critics. The group also has a large number of aquatic predators that live underwater and attack such insects as mosquitoes and midge larvae. These bugs come under a variety of descriptive names such as water boatmen, back swimmers, water-scorpions, giant water bugs, and water striders. Many miraculously find their way into backyard ponds. Another commonly heard (but not necessarily seen) homop in the garden is the cicada—an insect of Methuselahian legends. Some species of this group live upward of seventeen years, most of which is spent underground in the nymphal stage sucking on tree roots. Adults emerge, often in large numbers, and the females lay eggs in tree branches, which sometimes severely weakens the trees. Most of the summer, however, we hear the males buzzing in the trees, creating a perfectly wonderful din (if you like that sort of thing) in hopes of attracting a female or two.

Orthoptera: Cockroaches, Grasshoppers, Crickets, Katydids, Walking Sticks, Praying Mantids, and Rock Crawlers

This group is another difficult one for entomologists to organize. Its members are treated in from one to six different orders, seemingly depending on the time of day. Because the definition of these six groups is a complex task even for professionals, I think it best to treat all these creatures under one name. Each type is fairly distinct (does anyone not know a cockroach on sight?), and most of us can picture the majority of these insects, so that attempting to describe what defines each category is likely not needed. Rock crawlers, which are associated with glaciers, can be ignored for the purpose of this discussion.

The entire group is characterized, albeit loosely, by having two pairs of wings, the front pair of which is thickened and crisscrossed with many veins (creating many cells). All have chewing mouthparts and long antennae.

LIFE STAGES: Orthoptera, regardless of how one divides them, all undergo a simple metamorphosis with three life stages: egg, nymph, and adult.

NUMBER OF SPECIES: There are about 25,000 species in this order. The majority are grasshoppers, crickets, and katydids.

FEEDING HABITS: Orthoptera chew either plants or other insects or, in the case of cockroaches, whatever tastes best at the time. In fact, many plant-feeding Orthoptera consume dead organic matter as well as live plants. Grasshoppers are well-known chewers, especially in the Old World, where they periodically appear as plagues and swarms. Crickets and katydids are mostly plant feeders, but some will eat slow-moving insects, which perhaps they mistake for plants as they munch through their vegetarian dinners. Walking sticks are strictly plant feeders, and praying mantids are strictly carnivores.

GARDEN HABITS: The largest predatory insect the gardener is ever likely to see is the praying mantid. This insect, with its free-swiveling head and big, buggy eyes is one of the few insects that make almost lovable pets, especially for children. It is difficult not to like a

The lowly cockroach (Orthoptera),
shown here in its outdoor setting, is
one of the valuable scavengers and
elemental recyclers of the garden.

praying mantid, which is about the only reason a gardener should ever consider buying their egg cases to put in the garden. Mantids are great, voracious predators, but they will just as readily eat a butterfly or a bee as they will an errant beetle. By all means have them in the garden—they are great fun—but do not expect them to be of any use whatsoever except entertainment. You cannot train mantids to eat only what you want them to. Most Orthoptera are of little concern to the gardener, except when populations of grasshoppers and crickets (of all kinds) are extremely high. Then it is just a matter of too much of a normally innocuous thing. This order contains two of the great singers of summer, the katydids and crickets, and I, for one, cannot imagine a garden in the dog days of summer without its accompanying symphony orchestra.

Odonata: Dragonflies and Damselflies

Anyone who has pools of water in the garden is likely to attract these entirely likable creatures. The immature, or nymphal, stage is spent in the water, where the nymphs breathe through gills. In their search for water, adult dragonflies and damselflies investigate many shiny surfaces, including polished cars and oily spots in the road.

Odonata are distinguished by having two sets of membranous wings that are divided into hundreds of cells by a network of veins. In dragonflies, these wings are held outright when at rest, and in damselflies they are held upright over the thorax. Additionally, dragonflies and damselflies have large, faceted eyes that are used while hunting and that make it nearly impossible to sneak up on them when they are at rest. Odonata have elongate, needle-like bodies, which has inspired the common name darning needle or darner.

In Japan, dragonflies and damselflies are highly respected insects —its ancestral name, Land of Dragonflies, attests elegantly to this fact. Many hundreds of folk names exist for dragonflies, and there is even a Corporation for Consideration of Dragonflies and Nature. The Japanese, perhaps more right-thinking in some respects than us Westerners, also venerate many singing insects such as crickets and katydids (see chapter 13).

LIFE STAGES: Dragonflies and damselflies undergo a simple metamorphosis with three life stages: egg, nymph (spent entirely underwater), and adult.

NUMBER OF SPECIES: There are about 5000 described species of odonates in the world.

FEEDING HABITS: Dragonflies and damselflies are predators in every sense of the word. Nymphs hunt for prey underwater, which includes all other aquatic insects as well as the occasional small fish and tadpole. Adults are all predators, as well, preying on insects that they scoop out of the air, including mosquitoes.

Every garden should have water in it. This dragonfly (Odonata), ever the water seeker, displays why.

GARDEN HABITS: Anyone who has a pond should welcome drag-onflies and damselflies with open arms. They are not only beautiful, but also interesting to observe as they skim endlessly over the pond looking for food, mates, and places to lay their eggs. Some females simply wash the eggs off the end of their abdomen by dipping it into the water, others oviposit in shallow water by stabbing their eggs into the muck at the bottom, and others insert their eggs into vege-tation just below the water line. From a gardener's perspective, Odo-nata are almost perfect insects. The rumor that dragonflies will sew our eyes shut while we sleep is highly suspicious, and we gardeners need not fear such a fate. We're far too busy to take a nap.

Neuroptera: Lacewings (Aphidlions) and Antlions

This order is known to most gardeners, if at all, by the predatory green and brown lacewings and maybe secondarily by the antlions, or doodlebugs. There are other members of the group called snakeflies, owlflies, fishflies, alderflies, and dobsonflies, none of which, ironi-cally, are flies (that is, Diptera). Another group, the mantispids, look just like a green lacewing with a small praying mantid stuck on the front end. Some Neuroptera are fairly large: the dobsonfly, for exam-ple may have a wingspan of about 10–13 centimeters (4–5 inches).

Neuroptera have wings like dragonflies, except that they are folded rooflike over the back. Some antlion species have adults that are quite large and might superficially appear to be dragonflies, but they fly nothing at all like them—by comparison being rather feeble, floppy flyers.

LIFE STAGES: Neuroptera undergo a complete metamorphosis with four life stages: egg, larva, pupa, and adult.

NUMBER OF SPECIES: There are slightly fewer than 5000 described species of this order in the world.

FEEDING HABITS: Both adult and larval lacewings and antlions feed on other insects, and one species feeds on freshwater sponges. Adults have chewing mouthparts, but nymphs often have long, sickle-shaped jaws that are modified (along with another mouth-part) for grabbing, holding, and sucking. The well-known, green lacewing is a voracious feeder on aphids, with both larval and adult stages taking prey. Antlions feed on any number of small insects that slide into their funnel-like pit traps in the sand. A few lacewings are aquatic as larvae and feed on other aquatic insects. Mantispids feed as larvae in spider egg sacs.

GARDEN HABITS: Along with ladybird beetles, green lacewings are darlings of the biological control set and may be purchased from several different sources in the United States. Because green lacewings are sold as eggs (not adults, as with ladybugs), they would seem to be a better investment for the dollar. When you release ladybugs they all fly away—home, presumably! Lacewing eggs hatch, and the larvae crawl to food or die. If aphids are nearby, the larvae will stay put and do their job. Other members of Neuroptera have a fairly regional distribution. Antlions and owlflies are mostly found in the southern parts of the country, and snakeflies in the western part. It is probably safe to say that the green lacewing is the most likely and important neurotperan that a gardener is likely to encounter.

Isoptera: Termites

All termites are colonial and live in social aggregates founded by a reproductive queen and king. Termite societies differ in two respects from the social Hymenoptera (the honey bees, bumble bees, yellow jackets, hornets, and ants). First, in termites the male and female founders (the king and queen) live the remainder of their lives together within a wooden labyrinth and mate whenever necessary; in polite hymenopteran society, the male dies after mating and the female lives without mating again. Second, termite societies consist of both males and females at all times; hymenopteran societies consist solely of females, until a new colony is founded and then males are produced for one purpose only—a short life but a merry one.

At some point in their lives, reproductive termites have two pairs of membranous, many-veined wings, much like dragonflies, which are folded flat over the body. These wings are shed after mating and before establishing a new colony. Rarely does the gardener see the great swarms of winged male and female termites, which pair off to form new colonies, but swarms happen and so do new termite colonies, as some folks learn to their horror. About the only insect that a termite could be confused with is an ant—they are sometimes called white ants—and the differences are not so difficult to recognize. Termites are soft bodied, somewhat flattened, whitish in color, and have the abdomen broadly joined to the thorax. In contrast, ants are hard bodied; somewhat narrowed from side to side; black, red, or tan in color; and have the abdomen joined to the thorax by a narrow petiole.

LIFE STAGES: Termites undergo a simple metamorphosis with three life stages: egg, nymph, and adult.

NUMBER OF SPECIES: There are slightly fewer than 2000 species of termites in the world.

FEEDING HABITS: Termites eat many things, including grasses, twigs, leaves, plant litter, and animal droppings—that is, detritus of various sorts. Some filter the humus out of dirt. We are most familiar with termites as consumers of dead plant material, most notably trees as well as fences, porches, and houses made from trees. Termites have an enzyme that breaks down cellulose, which is the building material of plants. Some termites make this enzyme themselves, but others have a gut full of living protozoans that consume cellulose and help break it down for the termite to use as nutrition.

GARDEN HABITS: Termites serve a valuable function in the garden, that of recycler. Were it not for the fact that they sometimes get carried away—right into the house—we would not even know they were in our gardens consuming the various things that fall down and lay down when plants break apart or die. I will have more to say about these sorts of insects in chapter 5.

Dermaptera: Earwigs

I suppose an award might go to this order for having the most incomprehensible common name of any insect. Even if it were true, as sometimes written, that the name was derived from an old superstition that earwigs enter people's ears, the part about the wig is still open to conjecture. Do they enter only the ears of people who wear wigs? Another explanation is that the hindwing (folded under the short forewing) looks like an ear and thus this insect has an "earwing." Whatever the derivation of the name, in the eighteenth century earwigs were dried, pulverized, and mixed with hare urine as a medicine to help treat deafness. Whenever folks opine about the "good old days," I offer this remedy as a countermeasure.

Earwigs are recognized by the reduced forewings, which are leathery and partially cover the hindwings that jut slightly from under them. They also have forceplike pinchers at the end of the abdomen that are used for self-protection.

LIFE STAGES: Earwigs undergo a simple metamorphosis with three life stages: egg, nymph, and adult.

NUMBER OF SPECIES: There are about 1200 species of earwigs in the world.

FEEDING HABITS: A few species of earwigs feed on the skin of bats and rodents, but most species are considered to be scavengers that feed on bits and pieces of plant and animal material. The earwigs commonly found in our gardens have consistently been implicated in feeding on vegetable crops, fruit trees, and ornamental plants, but this may or may not be entirely true, as discussed below. Earwigs are nocturnal feeders and spend the day in protected locations. Thus, they fall victim to the rolled up newspapers that gardeners sometimes use to entrap them.

GARDEN HABITS: Earwigs have received much bad press in the past, but it is now uncertain whether they deserve it. There is some evidence that earwigs may be predators and scavengers, feeding on other insects as well as detritus found in the garden. It is possible that the obsessively tidy habits of gardeners may be one reason the insect has such a bad reputation (for details, see chapter 7). Another odd bit of information about earwigs is that females have the uncharacteristically motherly habit (in insects, anyway) of protecting their batches of eggs until they hatch. Does that sound like an insect that should be condemned without some additional thought?

Thysanoptera: Thrips

In yet another odd bit of naming, the common name of thrips is both singular and plural: one thrips is as good as a hundred thrips. Thrips are on the small side, rarely being more than a couple of millimeters (about ⅛ inch) in length. They are rarely seen, even by the dedicated gardener.

Thrips are distinguished by having two pairs of membranous wings, which are long, thin, and fringed with hairs. In case this is not enough information to make an identification, thrips are the only insects that do not have a right mandible—always a good thing to remember when the party gets a little dull.

LIFE STAGES: Thrips undergo a special sort of intermediate complete metamorphosis with five life stages: egg, larva, prepupa, pupa, and adult.

NUMBER OF SPECIES: There are about 5000 species of thrips in the world.

FEEDING HABITS: Most thrips rasp and suck plant juices, but a few are predators that suck juices from other insects, including other thrips species. Some are believed to feed on fungal spores.

GARDEN HABITS: Thrips are so tiny they often come and go before gardeners even knows what hit their plants. It takes a large population of rasping-sucking thrips to do any real damage. Sometimes they may be seen as elongate black slivers crawling over the surface of flowers. Some thrips transmit plant diseases, but generally the gardener is blissfully unaware of the entire problem.

Lesser-Known Orders

There are several other orders of insects that might turn up in the garden, and there are some that the gardener is likely to have never heard of and will never see (hopefully—lice come to mind here). In either case, the gardener conceivably will not care that these insects are in the garden or might pass through it, but in the name of completeness, I summarize each of the remaining orders for those who enjoy knowing things for no apparent reason. The orders are presented in several categories for your remembering pleasure.

Orders Seen Occasionally

SIPHONAPTERA (Fleas)—Flea species number about 2300. All are parasites of vertebrates from which they suck blood. Fleas undergo a complete metamorphosis and lay eggs either on the host or in the nest or sleeping area of the host. Flea eggs are often shed from dogs and cats into their sleeping areas in the house or onto the carpets. These eggs hatch into larvae that feed on bits of dander and organic debris. The larvae pupate and change to adults, which must have a blood meal to develop more eggs. In the case of dogs and cats, fleas can cycle in the home or in the garden, they're not too fussy. There are separate dog- and cat-flea species, but in a pinch, either species will attack both hosts as well as humans.

THYSANURA (Silverfish and Firebrats, or Bristletails)—There are 350 species of silverfish in the world. These are the insects sometimes seen in the cupboard or sink. They are flat, covered with silvery scales, and appear to have two or three long feelers sticking out one end and two out the other. This group has an inordinate fondness for starch and therefore may be found around books, bindings, labels, clothing, wallpaper, and anything else with starch in it. Silverfish prefer moist areas such as basements and sinks, whereas firebrats prefer warm areas such as furnace rooms and kitchens.

PSOCOPTERA (Psocids, or Bark and Book Lice)—There are about 2500 species of psocids. These insects are tiny. Book lice may be found on the unkempt desk or library, where they feed on bits of mold, fungi, dead insects, and glue used in book binding. If you move a paper on your desk and see a tiny, pale speck moving about as well, it is likely a book louse. Bark lice are found on bark and twigs, but you will likely never see them.

COLLEMBOLA (Springtails)—This order contains some 6000 species, which may or may not be insects (we are still working on that one). Whatever they might be, they are wingless and bounce around on the ground in the garden by the thousands. They are about the size of half a pinhead, feed on dead organic matter, and live in moist, shady gardens. If you want to see some, put a dash of detergent in a bowl of water and set the bowl flush with the ground. Come back a few days later. If you see something that looks like pepper, those are Collembola.

MECOPTERA (Scorpionflies)—There are about 400 species of scorpionflies worldwide. The larvae of this order live at or beneath the soil surface, where they feed on dead insects. The males of some species have a tail end that looks like the stinger of a scorpion, but none of these insects can sting or bite. Adults are predators on other insects, although some prefer to collect their prey already dead. Some scorpionflies are adept at stealing insects trapped in spider webs. Some males catch living prey with their hindlegs and then present this to a female, who is mated while she dines—a daringly efficient dating technique, when you think about it.

Orders Associated with Water

One insect order, the Odonata (dragonflies and damselflies), is commonly associated with garden ponds, but there are a few orders that gardeners are likely to find if they live near running water or a lake.

EPHEMEROPTERA (Mayflies)—This order contains some 2000 species. Their immature stages live in freshwater ponds, lakes, streams, or rivers, and feed on plants or algae. The adults fly about for a day or so, mate, lay eggs, and die. Adult mayflies have no mouthparts and thus do not feed. If you have a pond, you will sometimes find the shed nymphal skins near lights and even in your house (as I do). Mayflies are unique among insects in being able to fly before they are true adults.

PLECOPTERA (Stoneflies)—This order has about 1700 species. All stonefly nymphs live underwater in ponds, lakes, streams, and rivers; they either eat aquatic vegetation or are predators. The adults may not feed. Stonefly adults seldom stray far from aquatic environments.

TRICHOPTERA (Caddisflies)—There are some 7000 species of caddisflies worldwide. All caddisfly larvae live in streams, ponds, or lakes, where most species create little tubes of silk, embedded with sand and debris, in which they live. The head end of the larva sticks out one end of the tube. Some caddisfly larvae feed on aquatic plants and others on insects, but generally the adults do not feed at all.

Orders Hopefully Never Seen
The true lice (as opposed to insects with names such as plant lice [aphids] or book lice [psocids]) are placed either in a single order, Phthiraptera, or in two separate orders with more easily pronounced names. Either way, no one wants to find lice anywhere near their garden, because if they do, the lice may well be found on the gardener.

MALLOPHAGA (Chewing Lice)—About 2600 species of chewing lice are found in the world. These insects are parasites on mammals and birds, where they feed on hair, feathers, or skin. They are irritating and, if abundant, can cause an animal to become emaciated. Fortunately, not one of these lice attacks humans.

ANOPLURA (Sucking Lice)—Unfortunately, some of these *do* attack humans. There are just over 200 species of sucking lice. These insects are wingless and cling their entire life to a mammalian host. Two species attack humans: the body louse, which has a form that prefers heads and one that prefers other parts; and the pubic louse, which definitely knows where it wants to be.

Orders Rarely Seen Even by Entomologists
EMBIOPTERA (Webspinners)—This group, also called Embiidina, is composed of about 150 species, all of which spin silk from their front feet. They are less than the length of a grain of rice and live in colonies in silken tubes, which are built under rocks and debris and among mosses and lichens. Webspinners feed on dead plant material.

ZORAPTERA (Zorapterans)—There are some twenty species of zorapterans in the world. Besides being rare, they are so tiny as to be invisible. In spite of their size, they are predators on species of mites and insects that live in rotting logs and under bark. Zorapterans seem to be fond of hunting for prey in rotting sawdust.

STREPSIPTERA (Twisted-winged Parasites)—This group contains about 300 rare species found worldwide. All are internal parasites of other insects, especially bees and leafhoppers. Some attack crickets and even silverfish. Often a portion of the parasite can be seen sticking out, tumorlike, from between abdominal segments of their host.

PROTURA (Proturans) and Diplura (Diplurans)—These creatures, of which there are about 100 species worldwide, are generally less than 1 millimeter (0.04 inch) in length and feed on decomposing organic matter. Proturans and diplurans are considered rare and may not even be insects. I have never seen either group, except in an entomology class many years ago.

Well, if you have managed to stick with it this far, the tour of insects for insects' sake is over. You have learned about as much as a first-year entomology student, but they must go a few steps further into the families of each order, which takes some real time to digest. Students must also learn morphology, function, development, physiology, systematics, economics, ecology, toxicology, medical entomology, forensic entomology, and a number of additional specialized categories. As gardeners, of course, we do not need to know all these things, but we do need to know some of them. So, take a break, pour yourself a drink, and prepare to turn the page.

Then, again, maybe tomorrow would be a better day to continue.

Garden Arthropods

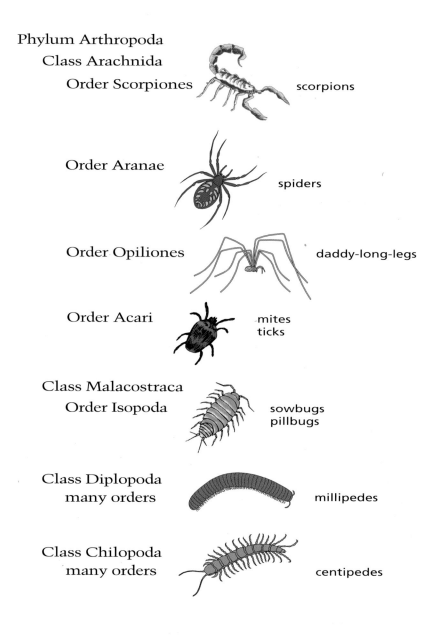

Phylum Arthropoda
  Class Arachnida
    Order Scorpiones   scorpions

    Order Aranae   spiders

    Order Opiliones   daddy-long-legs

    Order Acari   mites
                      ticks

  Class Malacostraca
    Order Isopoda   sowbugs
                      pillbugs

  Class Diplopoda
    many orders   millipedes

  Class Chilopoda
    many orders   centipedes

Class Insecta

Order Collembola 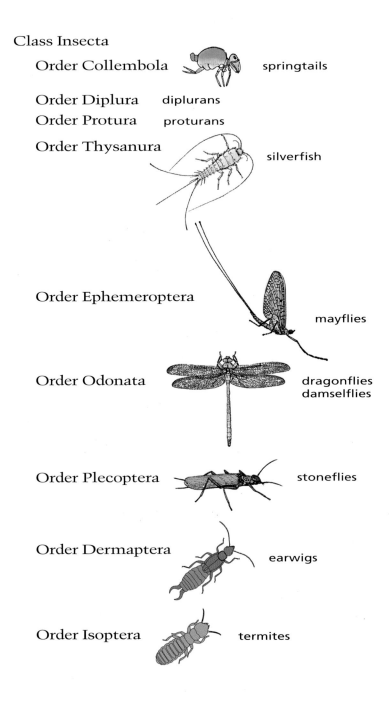 springtails

Order Diplura   diplurans
Order Protura   proturans

Order Thysanura   silverfish

Order Ephemeroptera

                                             mayflies

Order Odonata   dragonflies
                                             damselflies

Order Plecoptera   stoneflies

Order Dermaptera   earwigs

Order Isoptera   termites

Order Orthoptera
The following five groups
are treated as orders by
some specialists

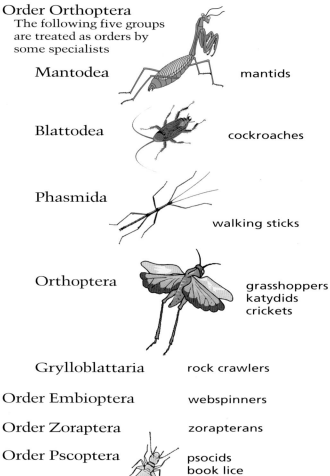

Mantodea   mantids

Blattodea   cockroaches

Phasmida

walking sticks

Orthoptera   grasshoppers
katydids
crickets

Grylloblattaria   rock crawlers

Order Embioptera   webspinners

Order Zoraptera   zorapterans

Order Pscoptera   psocids
book lice
bark lice

Order Phthiraptera   lice
The following two groups
are treated as orders by
some specialists

Mallophaga   chewing lice
Anoplura   sucking lice

Order Strepsiptera   twisted-winged parasites

Order Mecoptera   scorpionflies

Class Insecta

Order Siphonaptera 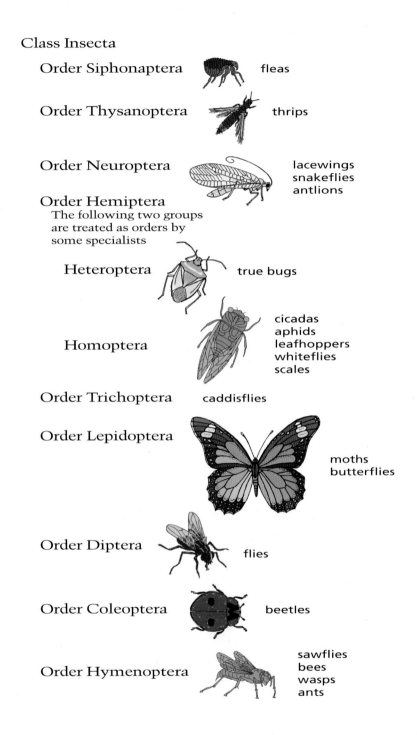 fleas

Order Thysanoptera thrips

Order Neuroptera lacewings
snakeflies
antlions

Order Hemiptera
The following two groups
are treated as orders by
some specialists

  Heteroptera true bugs

  Homoptera cicadas
aphids
leafhoppers
whiteflies
scales

Order Trichoptera caddisflies

Order Lepidoptera

moths
butterflies

Order Diptera flies

Order Coleoptera beetles

Order Hymenoptera sawflies
bees
wasps
ants

3

Some Basics of Insect Development

Insects are about as different from humans as is inhumanly possible. The biggest difference (if you ignore brain size) is that insects appear to be constructed inside out: their skeleton is on the outside. All an insect's internal body parts are encased within this skeleton, which is what we would normally think of as a skin. Technically speaking, we call the insect skin an exoskeleton. Because there are more insects than any other life-form, it must be considered the normal condition in nature for a skeleton to be on the outside. In other words, it is we who are relatively inside out, not the insects. These creatures have done quite nicely for 400 million years marching about with their rigid skeletons holding things together. We humans should be so well off.

Now, imagine if our own skin acted as a skeleton. We would be as rigid as the tin man in Oz, and we would be rusted tight as well because our skin is not composed of interfitting sections (that is, segments) as was the tin man's. Our skin is continuous. To hold an insect together and allow mobility, an external skeleton must be composed of loosely fitted segments that can move, or articulate, with each other. But such an arrangement poses several mechanical problems and one incomprehensibly complex problem as well. First, there would have to be some sort of membrane between the segments to keep all the inside stuff from leaking out. Second, this membrane would have to be flexible enough to allow some degree of motion. Because the tin man had no insides, not even a heart, leakage was not a concern for him. It is not a problem for insects, either, because each segment is joined to the next by a tough, flexible membrane, which allows a great deal of motion and flexibility—as any dragonfly would convince the most ardent skeptic.

The complex problem presented by an exoskeleton concerns development: how does the body become larger or change? Obviously, insects grow, otherwise they would never be larger than the eggs from which they emerged. Baby grasshoppers, which hatch at the size of a BB and are apparently wingless, mature into adult grasshoppers that can fly hundreds or thousands of miles. Even more spectacular, monarch butterflies grow as caterpillars and miraculously change into something they never appeared to be—a winged butterfly. It is amazing that a creature appearing to be nothing more than a bunch of stiff plates can grow or change into something it never was, but it happens trillions of times a day.

Insects must also reproduce their kind, which is basically what development is all about. Whereas some humans can imagine living only to produce great works of art, not just endless numbers of children, there is seemingly little reason for an insect to live unless it produces more insects. Along with the concept of reproduction comes the related necessity of individual contact, what we might call social interactions. Insects vary in societal mores all the way from reproduction without mates to reproductive cannibalism to highly complex monarchical societies. We will explore this aspect of development last.

Growth

All insects grow and change, but some do more with the process than others. There are three basic developmental patterns that reflect this, all of which involve the process of metamorphosis. One type of metamorphosis is simple, or incomplete. The second is complete metamorphosis. A third category, created to explain a few insects that do not fit either of the main two categories, is intermediate metamorphosis. Metamorphosis, of any sort, is a serious physio-logical change.

An insect's exoskeleton is composed of two parts. The external part is the cuticle, or cuticular layer (a structure similar to human hair), and the inner part is the epidermal (cellular) layer. When an in-sect grows, the epidermal layer secretes a new, soft cuticular layer that resides just under the old cuticle. A molting fluid is secreted between the two layers that causes the old, stiff cuticle to split. It then breaks away from the new cuticular layer, which is soft enough to expand. Soon the soft, new cuticle hardens, and the insect is again wrapped in a relatively rigid exoskeleton. In this way, an insect grad-ually shifts in size to become larger.

The process of growth in insects is incremental, or stepwise, not gradual as in vertebrates like birds, fishes, and humans. Stated sim-ply, insects have a hardened stage interrupted by a changing stage that is called a molt. The stage of life between molts, that is, what we normally call the immature insect, is termed an instar. In practi-cal terms, after it hatches from an egg, a grasshopper may go through four or five instars before it becomes an adult. Similarly, a monarch butterfly caterpillar goes through several instars before it becomes an adult. At the point of adulthood, no more molting takes place. The grasshopper and butterfly are essentially as big as they are going to get. Adult insects cannot grow larger, with the exception that the abdomens of female insects can swell as eggs develop.

The physical process of increasing in size is only one aspect of metamorphosis. Depending on the insect, there are minor or major changes in body shape and body parts and behavioral modifications such as flying, hunting, silk spinning, or mating. These changes de-pend on the type of metamorphosis involved: simple, complete, or intermediate.

Simple Metamorphosis

This type of development is also termed gradual or incomplete meta-morphosis. In general, insects of this type have three distinct life stages: egg, nymph, and adult. The name indicates that the transformation is neither as complex nor as complete as insects with complete metamorphosis.

In the grasshopper, an adult female lays an egg. This egg hatches and the first instar nymph emerges looking fairly much like a small grasshopper except that it appears to have no wings. What is not apparent to most observers, however, is that there are tiny, external pads where the wings will eventually be. One indicator of simple metamorphosis is that wings develop externally. As the insect grows, through the process of molting, the wings increase in size as do the pads that cover them. At the end of the grasshopper's final instar, it molts and four functional wings appear as the adult emerges.

In some cases, aphids being a common example, the adult may not appear any different than the nymph except in size. This is because only some adult aphids have wings. They sometimes have a generation or two without wings, then begin to produce winged forms. In this case, when there are wings, they again develop externally.

Most worrisome, perhaps, to the tidy minded—as I tend to be—is the case of aquatic insects that undergo a simple metamorphosis. The nymphs of dragonflies live underwater and look nothing at all like adults. In this respect, you might think that their metamorphosis is like a butterfly (that is, complete): the adult transforms from something that looks nothing like it. In the case of dragonflies, however, the last instar nymph leaves the water, crawls up a plant stem, splits it exoskeleton, and out pops an adult. Mayflies are even more daring. Their nymphs are aquatic, too, but they can fly in the last nymphal stage because the external wings are developed. Mayflies emerge from the water, fly to a resting point, rest for a while, then split the final nymphal skin and emerge as an adult.

Complete Metamorphosis

Insects with this type of development have four life stages: egg, larva, pupa, and adult. Such is the life of the typical butterfly or moth. There are several basic differences between complete and simple metamorphosis. In the former, the larval stage looks nothing at all like the adult stage, the wings develop internally (and are thus

When an insect undergoing simple metamorphosis molts, its outer skin (exoskeleton) slips off to reveal a larger, more developed replica. This grasshopper appears to be studying a past life, now held in its front legs.

never seen), and there is an interval of rest between the larval and adult stage, the pupal stage, in which the larva is reorganized into the adult.

The pupal stage of insects occurs in several different forms depending on the group. The usual type we think of is perhaps the chrysalid of the monarch butterfly. Here the pupal case is smooth and hangs from a pedicel of silken filaments. Butterflies generally have a bare pupal case, although most are cryptically colored and hidden from view, unlike the monarch's chrysalid. In moths, the caterpillar usually finds a protected spot, spins a silken cocoon, and then pupates within it. Lacewings, fleas, bees, and ants pupate in such silken cocoons. Many beetles and parasitic wasps, especially those that live in protected places such as plants stems, tree trunks, or soil, simply change from the larval to the pupal stage in the spot where they finished feeding. In many of the common flies, a pupal case is formed by the hardening of the last larval skin, which is given the special name *puparium.*

Numerically speaking, most insect species have complete metamorphosis. Just as butterflies and moths have larvae, so do bees, wasps, ants, beetles, flies, fleas, lacewings, and many others.

Intermediate Metamorphosis
A few insects develop in ways that do not fit neatly into the above two classes. Male scale insects, whiteflies, and thrips are the major culprits. All are classified as undergoing intermediate metamorphosis because during early stages some wing development is internal (as in complete metamorphosis) and later it is external (as in simple metamorphosis) and there is an inactive stage that precedes the adult.

Insect Size
To some degree, an insect's ultimate size is determined by its nymphal or larval nourishment. Caterpillars that consume more food may be larger than siblings that eat less food, and the adult will be relatively larger as well. But the adult itself cannot grow larger or smal-

An insect with complete metamorphosis hatches from an egg and
appears nothing like the adult it
will eventually become—in this
case, a giant silk moth.

Insects with complete metamorphosis change from larva to pupa to adult. In this photograph of an adult butterfly, the spiny, black, larval skin can be seen at the base of the pupal case from which the butterfly just emerged. The transformation from larva to adult is complete.

ler. Size is a relative thing, of course. An adult parasitic wasp the size of a rice grain can only vary in size within a certain range because its growth is genetically regulated, in spite of the quality and quantity of nourishment available to the larva.

Size is obviously variable between insect species. The smallest insect known is a parasitic wasp that feeds as a larva inside the egg of a bark louse. Two to four adult wasps emerge from each egg. A male wasp may be less than 0.15 millimeters (0.006 inch) long. At the other extreme, a Bornean walking stick may have a body length of 322 millimeters (12.9 inches), but with its legs fully extended it reaches 500 millimeters (20 inches). The largest insect ever found is a fossilized dragonfly with a wingspan of 700 millimeters (29 inches).

Because of the simplistic way (relative to vertebrates) that insects develop, there is a physical limit to the size that insects can possibly grow. This is determined by the strength of the exoskeleton versus the weight of an insect's body. The larger the body, the less structural strength it has. Fortunately for all of us, the Hollywood notion of a radiation-induced, giant cockroach that terrorizes Los Angeles is physically impossible. Such an insect would collapse under its own weight.

Reproduction

As all gardeners know, insects are adept at reproducing themselves. One reason that some species are so successful is that they create huge numbers of offspring, often in very short periods of time. Aphids come to mind, but aphids are atypical in many respects and should be thought of as oddballs of the insect world, as we shall see.

Most insects reproduce sexually, or what we think of as the normal way—a male fertilizes a female. (Oddly enough, this type of reproduction is often called bisexual to indicate that two sexes are involved in the process.) Because there are so many different species of insects, several kinds of reproduction have evolved that might be considered abnormal, at least from our point of view, but some are quite common, as we shall see. A perfect example to begin with is also one of our garden friends, the aphid.

Some insects, such as these true bugs, appear to be too embarrassed to face each other during the mating process.

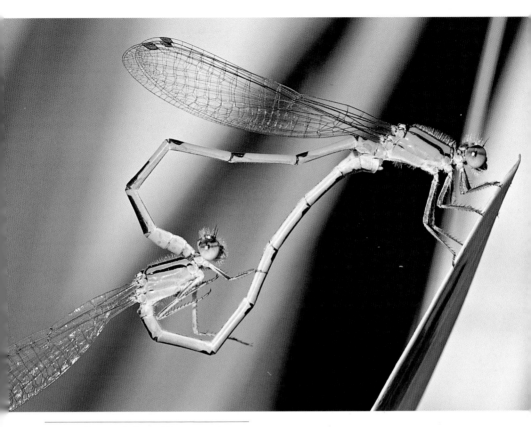

Dragonflies and damselflies mate in such a way as to provoke awe. In this photograph, a male damselfly (blue) has already transferred sperm from the tip of his abdomen to a special area at the base of his abdomen. The female (white) attaches the tip of her abdomen to the base of the male's abdomen to receive the sperm and is grasped at the top of the thorax (behind the head) by the apex of the male's abdomen.

These leaf-feeding beetles, in addition to being impossibly colored, are mating in the common insect way, male on top. An insect's reproductive organs are generally located at the apex of the abdomen.

Parthenogenesis

Parthenogenesis is the process of egg development without fertilization, and aphids are the queens of this biological art form. During a portion of their life, aphids produce nothing but females, which definitely precludes the notion of fertilization. They do this through the process of viviparity (birthing live young). A typical aphid life cycle goes something like this: overwintering eggs hatch in the spring; wingless females give birth to live young for several generations; when the environment becomes overcrowded, winged individuals arise and fly to a new host plant (or sometimes a different species of host plant); wingless females are produced again; these give birth to live young for several generations; winged individuals arise and fly to new host plants (or sometimes back to the original host plant); a generation of both males and females appears (the bisexual generation); these mate and the female lays eggs that overwinter. The production of wingless, parthenogenetic females is a summer phenomenon. In areas where it is consistently warm, no males may be produced for years.

Although aphids are among the great matriarchs of parthenogenesis, wasps, bees, and ants are the greatest proponents of the game. In all of these insects, females have some ability to control the sex of their offspring. This is called haplodiploidy. Under this arrangement, fertilized eggs develop into females and unfertilized eggs (that is, parthenogenetic reproduction) develop as males. This procedure is carried to an extreme in the social wasps, bees, and ants, in which the founding female (queen) is mated one or more times and then never mates again. She dispenses fertilized eggs during her lifetime, thus creating an all-female society, until such time as the token males are needed to fertilize new queens, and then the old queen lays unfertilized eggs. It is a highly efficient system, to say the least. In one ant species, a single queen apparently can store up to 7 million sperm to be used over a ten-year period.

Some wasps, bees, and ants also practice a specialized form of parthenogenesis involving alternating generations. This takes the concept of the aphid alternation of wingless and winged forms a bit further. A number of gall-forming wasps, or cynipids, have a partheno-genetic (uniparental or unisexual) form that has no males. These wasps lay eggs in plant tissue, especially oaks, from which both male and female wasps are produced. These bisexual offspring produce galls that look nothing like the gall of the unisexual parent wasp, nor do the wasps themselves bear any resemblance to their mothers. The offspring have often been classified as different species or even genera because the alternate generations have not been associated by rearing studies. The bisexual wasps mate and produce the unisex-ual, female wasp generation.

Parthenogenesis is not common in other groups of insects, but may be found in some beetles (for example, white-fringed beetles, where males are unknown), some thrips (males are rare or unknown in some species), and some walking sticks.

Viviparity

Viviparity is the condition in which eggs are retained in an insect's body, then hatch and enter the world as functional immature in-sects—live birth, if you will. Not only are aphids parthenogenetic, they are also viviparous. As noted above, adult females produce im-mature aphid nymphs within their bodies. Thus, we know that in-sects with simple metamorphosis can be viviparous. A few insects with complete metamorphosis can do the same. Some adult flies deposit fully grown larvae instead of eggs. Viviparity is also found in some cockroaches and scales.

Paedogenesis is a special sort of viviparity. This occurs when an immature stage, not the adult, gives birth, although birth is not quite the correct term. Some cecidomyiid fly and fungus gnat larvae produce live young, but in doing so the larva itself is killed. This bizarre form of reproduction is rare and is known so far only in a few flies and a beetle.

Polyembryony

A most bizarre case of reproduction is found in polyembryony, in which a single egg divides internally into a number of embryos, as if six chickens hatched from one egg. Certainly, this is an extremely efficient way to reproduce one's kind. Insects, of course, take any good idea to the extreme, and in a few species of parasitic wasps, a single egg may develop into as many as 1000 individuals within a host. Claims of up to 2500 individuals have been made, but this might be stretching the truth a bit.

Hermaphroditism

Finally, we come to a single case of extreme self-devotion. In hermaphrodites, the sex glands of both sexes are combined in a single individual. I know of only one example of this mode of reproduction in insects, the cottony cushion scale, which is found on citrus trees in California. Incidentally, the cottony cushion scale was the first modern (in the 1870s) example of biological control, the use of one organism to destroy another. In this case, it was the vedalia beetle introduced into California from Australia that saved the citrus industry.

Social Interaction

A hermaphrodite obviously needs little contact to keep itself amused, and the parthenogenetically able seem to have some choice in determining their own lives, at least at one point or another. But if we ignore the sorts of insects that do not need mates, we are confronted by the enormous world of sexual union. We will now look at associations between insects as a guide, in part, to their development and their ways of living in the world and our gardens. The gradation from completely self-absorbed to solitary to gregarious and finally to socially complicated is a long and tenuous one. In truth, most insects are considered to have a solitary disposition, getting together only at an appropriate time for a single purpose, conjugation. A number of highly successful insects, such as termites and ants, have given way to completely social lives. But, as always with insects, there are many odds and ends that defy neat categorization.

Solitary Insects

The majority of insects are solitary. They hunt for food alone, eat alone, sleep alone, and mate only when urged to do so by hormonal instincts. Almost any moth, butterfly, fly, beetle, parasitic wasp, dragonfly, or flea fits this category. Sometimes clusters of insects are found, but these are often accidents of birth. For example, if several female grasshoppers lay their eggs in the same area, it may lead to the appearance of gregariousness, but it may be just coincidence. (Sometimes, however, these coincidences turn into a sort of organized gregarious frenzy, as we see under Insect Aggregations below.) To reproduce, solitary insects must either be parthenogenetic or be able to find mates exactly when needed. Insects are very good at the former, as we have already seen, and they are even better at the latter.

To locate one another, one or both mates must initially recognize the other, and it would be a good thing if the attraction was species specific. It would do little good, for example, if a female moth attracted 300 different species of male moths—she would be auditioning them all night and never have a chance to lay eggs. In insects, methods of mate recognition are both diverse and highly selective.

Chemicals are one of the most specific signals used in the mate attraction ritual. These chemicals, given off by females (usually) or males (rarely), are called pheromones. They are unique to each species and highly attractive from relatively long distances. The signals are picked up by various chemical receptors (chemoreceptors) located on the male antenna. These pheromones can be extremely potent: there are reports of marked and released males being attracted to female pheromone traps from distances of four, eight, and even twenty-three miles away.

Male bumble bees are among the few males that use pheromones. A male will mark a series of perches with secretions from the mandibular gland (located in the mouth region). Females are attracted to these perches and wait for a male to return. In contrast, the males are busy marking perches and revisiting the old, marked ones much like a trapper with his lines.

In addition to helping insects recognize mates, there are other pheromones that organize basic types of communication. Some bark beetles, for instance, use pheromones as an attractant that causes both males and females to aggregate at a tree. Here the beetles mate,

Most insects are loners, seeking out members of their own kind simply to mate. This grouping of flower flies (syrphids) is most likely coincidental. A combination of numerous flies and little readily available nectar may have drawn them together momentarily.

and a combined army attacks a tree that fights back by exuding resinous sap. (There is victory in numbers.) Some female insects secrete pheromones to stop the mating process once mating has occurred. Also, some females, such as the apple maggot fly, secrete an anti-oviposition chemical that deters other females of the same species from using the same host.

Insects do not necessarily need chemicals to attract or recognize each other. Good old sight works sometimes, but it is a bit tricky because insects cannot see each other from miles away; they must be relatively close. Then there is always the problem of getting it right the first time, particularly if a predatory insect is dealing with a mate that would just as soon kill him (or her) as not. Thus have evolved splendid courtship rituals, during which certain movements must be made in particular sequences or tragedy will strike. We have all heard of the male mantid that must first get close enough to his mate without being eaten, court her, and then as soon as he does the deed, loses his head for all his trouble. Although this behavior may be exaggerated by an overdramatic press, a male mantid must certainly be careful when approaching his intended mate.

One group of insects communicates by flashes, not of insight, but of light. Fireflies, or lightning bugs, are beetles that have the ability to illuminate and darken their abdomens. Both males and females flash species-specific light signals, females while sitting and males while flying. The communicative flashes continue until a male recognizes a female of his own species. He then lands at her side to consummate all the flashing. Unfortunately, not all females play fair. Female species in one genus of fireflies flash their abdomens with signals of females of other species. When the excited male arrives to do his courting, the female kills and eats him. Life is not fair, we know this, but sometimes it becomes just a bit too vicious for my taste.

Another major way of mate communication is a bit more upbeat —in more ways than one—it is singing. In insects, singing involves a great many body parts, indeed, none of which has anything to do with the mouth. Some such as crickets and katydids sing (actually they stridulate) by rubbing their wings together. Others, such as cicadas have tympanum-like membranes stretched across circular rims on the sides of their first abdominal segments. These timpani are vibrated by muscles attached to the inside of the membrane and amplified by the largely hollow abdominal cavity. Some beetles, called bessbugs, rub their hindwings over a special rasp on the upper part of the abdomen. This rasping, courting song is performed by both sexes and can last more than twelve hours, at which point mating takes place, simply to stop all the racket. Well, presumably, at least.

Social Insects

The opposite of a solitary lifestyle is a social one. Social insects, as defined by some, are insects that cooperatively care for their offspring. Others say that there must be a division of labor for an insect group to be social, even in its most primitive expression. These two criteria, somewhat simple in character, are considered indicators for the subsocial way of life. If a third criterion is added, that of overlapping generations living together, then an insect group is considered to have complex interactions and is eusocial.

Although subsocial behavior is known from several groups of insects, it is not very common. Some cockroaches, true bugs, and even earwigs are known to show a degree of motherly concern to their offspring, so we might allow them some social acumen. Several different beetles are also subsocial. In bessbugs, mentioned above, a male and female carve out tunnels in a dead, fallen tree. Here they live with their twenty to sixty larval offspring. The adults feed their larvae by providing chewed up wood tissue mixed with saliva. The adults and larvae communicate by means of squeaking sounds. This, at least, sounds like the beginnings of proper social behavior.

Eusocial behavior is supposedly found in several insect orders, but it is has reached its most advanced state only in two groups: Isoptera (termites) and Hymenoptera (wasps, bees, and ants). In these groups, sociality is carried to a most efficient extreme. Because individuals behave so (apparently) precisely and perform such specialized jobs, they often have been compared to the cells of a single, liv-

In a rarely seen event, a female bumble bee investigates the cell from which one of her sisters emerged. These cells are formed from wax generated by the bee. Bumble bees are eusocial, living in colonies often founded in abandoned mouse nests or sometimes bird nests in protected places.

ing organism. We need not go that far in eulogizing social insects, but certainly there is little in the insect world to rival the complexity of a honey bee colony with its wing-induced air conditioning, sex rationing, nectar-to-honey factory, queen production, corpse disposal, general housekeeping, and its remarkable dance language in which source, direction, and distance to nectar is communicated by a few circles and wiggles of the body accompanied by olfactory signals. Rhapsodic books have been written to the honey bee as the crowning glory of societal evolution, and deservedly so, at least in nonhumans.

Ant societies are perhaps second only to the honey bees' in complexity of communication, but they outshine all other social insects in diversity of nest construction. There are almost 9000 species of ants that range over most of the world, so you would expect a great diversity of lifestyles. Communication in ants is mostly through pheromones as opposed to any sort of dancing language. Scout ants communicate the direction of newly found food sources via a pheromone trail they lay down as they head directly back to their home nest from the food. (Theoretically this straight route home should be an "antline," not a "beeline.") Ants are unique in their ability to communicate with two other orders of insects—or at least to cooperate with them peacefully enough that both ant and non-ant benefit from the relationship. Ants protect aphids from parasitoids and predators and even build shelters for them. One species of ant is so protective of a root aphid that the aphid cannot exist without the ant's care. In return for protection, aphids provide honeydew, a sugary substance upon which the ants feed. Ants also care for the larvae of several butterfly species (blues and hairstreaks), which produce similar sweet rewards for their protectors. (Anyone wishing to know more about ants should read the introduction to *The Ants* by Hölldobler and Wilson [1990].)

As noted in chapter 2, termite societies differ from those of bees and ants in that males live within the colony on a permanent basis. If termites are not as architecturally savvy as ants (some experts argue the point) or as eruditely communicative as honey bees, they make up for it in having the most complex system of caste structure of any insect. In termites, both males and females are found in each caste. There is a primary reproductive caste, a secondary reproductive caste, a worker caste (sometimes of two kinds), and a soldier caste (sometimes of three kinds).

Above, I mentioned that pheromones (chemical signals) helped individuals find each other so that they could create more of their own kind. These were air-borne chemicals that could spread for great distances. In eusocial insects, pheromones are also used to maintain the social structure in this big, riotous assembly. Eusocial

Ants are true eusocial insects that care for their young. In this photograph, the adult ants are watching over silken pupal cases, each spun by an individual larva.

insects need to work together in some organized fashion or utter chaos would result. This communal group structure, whether ant, bee, or termite, is held together by pheromones through the process of trophallaxis, which is the exchange of nourishment from one individual to another. Thus, mutual feeding becomes the ultimate method of chemical exchange and control within social insect groups.

In honey bee colonies, for example, the queen secretes a fluid from her mandibular area called queen substance. This secretion is taken up by the bees that tend her and then spreads outward in an ever-increasing circle as each bee feeds other bees. Trophallaxis was demonstrated in one experiment by feeding radioactive phosphorus to 6 bees in a colony of 25,000. Within twenty-four hours, 15,000 bees were radioactive. This queen substance serves two purposes: to communicate to all the worker bees that the queen is "in the house" and to repress the sexual development of all workers in the colony (recall that all worker bees are females). Thus, the only reproductive female in a colony is the queen until something happens to her or the colony grows so large that there are not enough of her pheromones to go around. At that point, new queens arise and a swarm eventually ensues.

Insect societies are complicated—of this there is little doubt. The fact that so few groups display sociality speaks to the difficulty of coercing individuals to work together for a common purpose, even one as essential as procreation. Perhaps not so surprising, in light of both extremes of the solitary individual and communal society, is the number of insect species that exhibit some form of intermediate gregariousness, for lack of a better term: the ability to be together, but at the same time to be separate.

Insect Aggregations

Many insects end up together through no design of their own. They are creatures of circumstance. Adult female aphids, for example, simply plop their endless supply of offspring into the world and then live among them. In my opinion, this seems to make them eminently underqualified to be considered anything but lazy, but a few aphid species are claimed to be eusocial. (I still have my doubts.)

Many parasitic wasps insert more than one egg into a host, with the result that some hosts many contain dozens to thousands of wasp larvae inside them. These are no more social than rocks in a pile, but they represent aggregations by anyone's standard. Some gall-forming flies and wasps lay multiple eggs in a single location on the same plant host and the resulting galls may develop into multichambered swellings, each housing several to a dozen larvae. Inside the single gall, however, each larva is individually surrounded by plant tissue so that no larva ever touches another one. Externally the emerging adults seem to be gregarious, but internally they each inhabit their own world.

Some insects aggregate on plants as a result of having been born from the same cluster of eggs. Instead of spreading out immediately, as most insects do, the larvae or nymphs of these aggregators stay together for varying amounts of time. Some moth larvae stay together for several instars, then move away from each other. Tent caterpillars remain together their entire immature life, an attraction created by aggregation pheromones that induce larvae to build a massive web-nest in which they spend the day out of harm's way. At night, tent caterpillars forage out on the leaf-producing branches, where they feed under cover of darkness. Pine sawfly larvae congregate from eggs to pupae as they feed a branch at a time. More often, the gardener sees the denuded pine branch before a single larva is seen. Sawfly larvae gain some degree of protection in numbers because they are full of chemical compounds obtained from pine needles. The chances of more than one larva being attacked by a bird, for example, is slim.

Other aggregations are also accidents of birth. Some soil-nesting solitary bees and wasps build nests in the same region as their mother. Some permanent sandy areas, in the desert, for example, end up with hundreds or thousands of individuals nesting in the same relatively small area simply because nesting has occurred year after year without major interruption. A few females are always

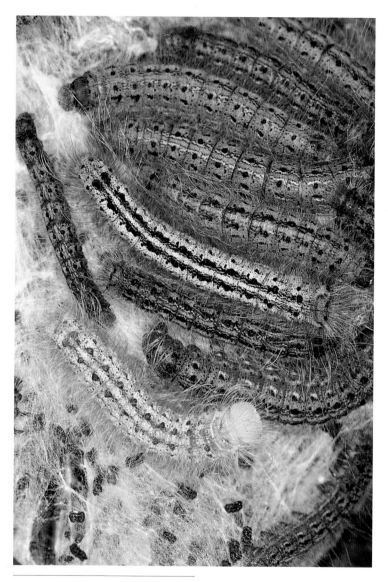

Western tent caterpillars cluster together, but have no real social interactions. Perhaps some form of protection is engendered by living in clusters.

Clusters of western tent caterpillar webs
are evident in this aspen grove long
after the larvae have finished feeding
and the adults have flown away.

leaving the area by accident (winds, getting lost, being chased away), and these may colonize new areas long distances from the original nest site. But, upon leaving their underground cell, many newly emergent females find a nearby area suitable for nesting and begin the cycle without straying too far.

Some insects congregate overnight in sleeping groups. This is true of some bees and predatory wasps. The monarch butterfly gathers in huge numbers in several overwintering spots. It is possible that aggregating in a group provides some degree of protection induced either by collective bad taste (monarchs), collective stinging (wasps), or just plain safety in numbers. There is less likelihood of an individual being killed in a huge aggregation. Some insects aggregate during hibernation. Most notable, perhaps, is the ladybird beetle, which overwinters in such huge congregations that it becomes economically feasible to collect and sell them to suppliers of biological control agents.

As can be seen from the basic examples outlined above, insects lead amply exciting lives. From the complexity of growth, to the perplexity of sex, and even the intensity of social interactions, insects are not creatures to be simply ignored. Perhaps through a better understanding of the formative aspects of insect life we might learn to appreciate them just a bit more before we bring our soles, or our chemicals, to bear on their fragile, apparently worthless bodies.

4

Survival

In chapter 2 we learned a bit about the orders of insects likely (or unlikely) to be found in the garden. Here we turn to the subject of how those insects live and in what sorts of habitats they will be found. As with many fundamental efforts in life, the answers we seek mainly depend on the questions we ask.

For example, if we want to know what insect just flew by us and landed on the rutabagas, we examine the insect (or remember its details) and try to deduce from its appearance what it is. We ask the basic question "What is that?" Often, however, we do not see an insect at all. We see a hole in the rutabaga leaf and say, "What did that?" The answers to both questions may be the same or they might not be, but the two questions are different and they require different types of knowledge to answer. One of the purposes of this book is to acquaint the reader with different ways of looking at insects, because no one perspective answers all the questions all the time. The truth is that insects, in all their explosive diversity, are so confounding that it often takes numerous questions and ways of looking at them to understand much of anything.

Feeding Mechanisms

Before I discuss feeding habits it is necessary to describe, in a simple way, the structural arrangement of an insect's feeding end, or what we would call a mouth. I say "in a simple way" because the mouth-parts of an insect are rather complex, with lots of different arrangements that can be seen only at a microscopic level. Recall that insects have their skeletons on the outside. For insects, eating is much the same as if we had our teeth and jaws outside our mouth, only it is more complicated. The structures that insects use to get their food into their mouth are derived from the legs of an ancestral, insectlike creature that had many segments each with paired legs. Thus, to put it simply and brutally, insects shove food into their oral cavity with three pairs of appendages that used to be legs. We have names for each of these, but probably the most recognizable term is *mandibles* for the biggest pair (what we think of as external jaws).

There are two elementary ways that these mouthparts are arranged in different insects. The jawlike mandibles are the simplest and are used for the basic sort of grinding, chewing, and tearing that we see in grasshoppers, caterpillars, and wasps. The second is a sucking mechanism, in which mouthparts are fused to form different sorts of tubes. Mosquitoes, for example, have a stiff, piercing tube that is well known for its blood-pumping ability. In contrast, butterflies have a soft, rolled up tube that siphons nectar from flowers. Flies have a sponging arrangement in which the ends of some mouthparts are expanded to sop up liquid, which is transported via a tube through the upper mouthparts.

How insects feed is one key element in identifying an insect we cannot see. If, for example, holes are cut in the leaves of plants, we can be certain that a chewing insect is the culprit (once slugs, snails, rabbits, and deer have been ruled out, of course). Alternatively, if leaves are covered with discolored spots, we are likely to be dealing with a piercing-sucking sort of insect. This is fairly simple-minded stuff, you might think (and rightly so), but sometimes we harried gardeners need a large variety of simple information to derive intelligent conclusions about what insects are doing in the garden.

To better understand insects, though, we need to take the mechanism of feeding a step further. As we saw in chapter 3, insects can have different feeding stages at different times in their lives. Those that undergo simple metamorphosis feed in the nymphal and adult stage, and those with complete metamorphosis feed in the larval and adult stage (the pupal stage does not feed). A couple of fine details can be drawn from each category that may help us understand how insects are feeding in the garden.

In insects with simple metamorphosis, for example, there is no great change between the nymphal and adult stages, so in essence whatever the nymph eats, the adult continues to eat, except more of it. Grasshoppers are a good example. The adult simply chews what it has chewed all its nymphal life. An aphid simply sucks what it has sucked throughout its nymphal life. Thus, simple metamorphosis pretty much equates to a fixed diet.

Insects with a complete metamorphosis, however, may be a bit more complicated because there are two feeding stages (larval and

Many insects have strong, opposing jaws with which they chew their food. This carabid, or ground, beetle is a hunter.

adult) interrupted by a nonfeeding, pupal stage. Thus, a caterpillar with jaws feeds on a cabbage plant, but when it emerges from the pupal stage, the adult butterfly feeds on nectar with a siphoning tube. Most mosquito larvae are filter feeders, drawing in bits of underwater debris and microorganisms, whereas most adult females

Adult moth and butterfly mouthparts are formed into a thin, flexible tube through which they suck nectar and sometimes dissolved pollen. The mouthparts of their larvae, however, are opposing jaws. Both adult and larval moths and butterflies are herbivores (with a few larval exceptions).

are blood suckers and the more mild-mannered males are content to sip nectar from flowers. Complete metamorphosis allows some insects, but not all, to alter their feeding methods, habits, and habitats during different parts of their lives. Thus, a single species, depending on its developmental stage, may occupy several different feeding positions in the habitat.

Feeding Types

Not all insects feed throughout their lives. Some adults have no mouthparts and are simply reproducers. Some stages, such as pupae, do not eat. Some insects can go into an inanimate state, either as larvae or pupae, for months or years and do not eat. However, at some time in the lives of all insects, they must obtain nourishment.

Eating is a simple process that becomes somewhat complex when one is pressed to explain. There are three main categories of consumption that we generally recognize: animals that eat meat are carnivores, those that eat plants are herbivores, and those that eat both are omnivores. Humans generally are thought of as omnivores, although some folks are vegetarians (of which there are degrees of classification). No human, I believe, survives on a strict regimen of meat alone, and so we probably cannot be considered pure carnivores. Insects, too, fall into these three categories. But, as with all things insect, there is a myriad of nuances.

Scavengers

Insects that specialize in feeding on decomposing life-forms or the by-products of these once-living forms are scavengers. Some insects specialize in eating dead vegetation and would consider a fresh Brussels sprout inedible. Other insects specialize in eating dead animals, with some dining only on certain stages of the decaying process. Yet other insects confine themselves to eating dung, for which they receive the ignominious name of scatophagous (or coprophagous) insects.

It is difficult to find a single insect order that is devoted entirely to scavenging. The cockroach, which is sometimes treated as a distinct order, comes as close as any. Perhaps another insect that comes to mind is the ant, a creature we see everywhere and apparently eating everything, but it is only part of the order Hymenoptera. There are so many species of ants that it is impossible to paint them all with a single stroke of the brush. About the only statement that safely can be made is that many ants are scavengers and the other species are herbivores, carnivores, and omnivores. Several insect orders are as equally ambiguous as the ants in their feeding habits.

Herbivores

Insects that feed on living plants are herbivores and their habit is herbivory. In chapter 6, I treat these sorts of insects in detail by defining and examining the various parts of the plants and the different insects that feed on them. Here, I examine only the basic ways in which insects attack plants.

The most readily observed method of plant feeding is the free-living insect that wanders about eating plant parts. Whether it is an adult Japanese beetle or a tomato hornworm larva, the insect is simply munching away as if its life depended on it—which it does. An insect does not have to wander about and munch, though, to fit in this category: it can simply sit and suck—like an aphid. There is nothing too complicated about insects that feed externally on plants. It is this sort of insect that often gets the gardener riled up to the point of a chemically dependent frenzy. However, there are other, more insidious plant feeders.

The borers, for example, are insect larvae that get into plant tissue and chew their way through the birch terminals or iris rhizomes before a gardener knows what bit them. If the insect does its boring within leaf tissue, we call it a leaf miner and can readily observe the traces of its mines as a bas relief in transparent leaf. Then there are the gall-forming insects that cause an increase in plant tissue and/or an increase in the size of plant cells surrounding a larva that feeds in the tissue. Sometimes the galls are induced by chemicals injected by an adult insect when she lays an egg. These galls appear as visible swellings in some part of the plant.

As we shall see in chapter 6, herbivores use the basic approaches just enumerated to attack just about every part of a plant imaginable.

Carnivores

Animals that eat other animals are carnivores. Sometimes, when they eat insects, they are given the specialized name entomophagous. That is, birds that eat bugs are entomophagous carnivores. Most carnivorous insects are entomophagous, but surprisingly there are a few insects that can kill and eat small lizards, birds, fish, tad-

Some insects, such as this weevil, appear to have long, piercing-sucking mouthparts. In reality, the elongation is part of the weevil's head and is equipped with jaws at its tip. Insects with piercing-sucking mouthparts cannot chew holes in plants as weevils do.

poles, and frogs. There are two main types of entomophagous insects, predators and parasitoids, each of which has a particular
method of attack and will be explored below. The dynamic and specific interactions between insects are treated more fully in chapter 7.
Parasitic insects are discussed in the last part of this section, not because they are carnivores, but because they confuse the issue a bit.

PREDATORS—This behavior is what we think of as hunting—animals that attack and kill other animals. Usually there is a caveat
that a predator must kill many prey to keep itself or its young alive.
The insect world is rife with examples of these killers, and the following will likely be known to most gardeners: lacewings, dragonflies, praying mantids, true bugs, ladybugs, syrphid flies, and yellow
jackets. Many predatory insects can attack at several life stages including nymphal and adult. Even a few moth larvae are known to
kill other insects for food.

Many solitary wasps and all social ones take the predatory nature
a bit further than the simple kill-and-eat-it strategy. Adult female
wasps hunt for prey, which they feed to their offspring. In the cases
of yellow jackets and hornets, the female chews up the prey and
feeds it directly to larvae much as a bird feeds its nestlings regurgi-
tated food. The solitary wasps, however, generally paralyze their
prey by stinging it, then place this food in a protected spot (inside a
burrow or twig, for example), lay an egg on it, and go away. The egg
hatches and the larva feeds on its provisions, which are all it is going
to get. This behavior approaches that of the parasitoid, which we
examine next.

PARASITOIDS—Although most gardeners have probably heard the
term *parasite,* and perhaps even used it in polite company, I doubt
many have heard the term *parasitoid.* Unfortunately, all but the
most discerning gardeners do not realize that they are using the former term incorrectly—at least entomologically speaking. The term
parasitoid is reserved for insects that lay their eggs on or in a host insect. (An explanation of the term parasite will be given in the next
section, simply to keep things from becoming too complicated here.)

Parasitoids do not gather up their hosts and stuff them someplace convenient, as do the predatory wasps just mentioned. In general, a female parasitoid finds the host she wants, inserts an egg into it, and flies away. In some cases the host continues to wander around, feed, and grow—totally unaware that bad things are going on inside. The egg (or eggs) of the parasitoid hatches, and the resultant larva either feeds as the host grows or it may wait until the host completes growth before it gobbles it up. Either way, the host eventually becomes supper for the parasitoid(s) it harbors.

Predatory robber flies have piercing-sucking mouthparts, which they insert into the bodies of any creature they can overcome, large or small. The robber fly (on top) has defeated a blister beetle and is piercing the soft membrane between the beetle's thorax and abdomen. As adults, robber flies are aerial predators. As larvae, they have jaws and feed on soil-inhabiting insects.

106

There are a million variations on this theme, but for now I'll just say that there are parasitoids that lay eggs inside the host (internal, or endoparasitoids) or outside the host (external, or ectoparasitoids) and these methods of attack depend, in part, on how protected the host is. A fat, free-living tomato hornworm sitting right out for the whole world to see almost requires that a parasitoid develop inside for the parasitoid's own protection. A stem-boring beetle larva, however, is protected enough so that its parasitoid can safely feed on it from the outside of its body (but inside the stem, of course).

Adult parasitoids are generally nectar feeders, although some do feed on hosts at the point of oviposition, where the host's body fluids may leak a bit. It is their progeny that do the flesh eating. In most cases, one host dies and that is the end of that. Technically speaking, these sorts of insects have a parasitoidic (not parasitic) form of behavior, but this term is so overly pedantic that in the broad scheme of real life, real people, including real entomologists, usually refer to parasitoidic insects simply as parasitic. I usually do so myself, wrong as I may be. For those of you who feel comfortable being overly pedantic, I present the ensuing discussion on true parasites simply to make your life complete.

PARASITES—In the entomological world, we are forced to deal with large numbers of insects and great diversities of behaviors. Because of this, we entomologists often end up splitting categories into fairly detailed subsections because it is the only way we can talk about something and really have the words mean what we say. So it is in the case of parasites and parasitoids.

Insects, such as mosquitoes, fleas, bedbugs, lice, and even some earwigs are called parasites because they feed externally upon a host animal for part or all of their life cycles. These hosts tend to be warm-blooded mammals or birds, and the feeding tends to center around draining various amounts of blood from the host's body. Some of these parasites, however, merely dine on dead, flaking skin and bits of sloughing stuff associated with a host's body.

Generally, mosquitoes, fleas, bedbugs, and lice do not kill their hosts simply by removing a little blood now and again. (They may deposit disease-bearing organisms into the blood, which might kill their host, but they do not kill directly.) These insects feed on many different individuals of the same host (and sometimes other species of host), but they are not predators—they do not kill. These true parasites are considered to have a parasitic form of behavior. With the exception of mosquitoes (and ticks, which are not insects, as you may recall), these are not the sorts of insects we gardeners worry much about in the garden.

If you need a way to remember the difference between a true par-

This braconid wasp has just emerged from its host caterpillar. Parasitic wasps (parasitoids) have jaws as both larvae and adults. Parasitoid larvae are entomophagous, or insect eaters, but most adults feed on nectar and pollen. When some female parasitoids puncture the host insect with their ovipositors (or egg-laying tube), they feed at the puncture site either before or after laying eggs.

asite and a parasitoid, it might be useful to remember that you would much rather be bitten or pierced by a parasite (for example, a mosquito) than eaten alive from the inside by a parasitoid—which fortunately will never happen. To set your mind at ease, I can assure you that no insect parasitoid has ever attacked a human being in any way, shape, or form. Nor can they sting, because they do not have stingers, as do the more advanced wasps, such as the hornet. There is nothing for humans to fear from parasitoids—not even fear itself. What appear to be stingers are simply egg-laying tubes, and parasitoids have no interest in humans whatsoever. Parasites, on

Many parasitoids emerge as adults from their host, but some emerge as larvae and pupate outside the host. The pink parasitoid wasp larva shown here emerged from its caterpillar host—now a mere shell of its former self—and is preparing to spin a cocoon.

the other hand, often feed on humans with great glee—or at least as much glee as can be interpreted from a louse's facial expressions.

Habitat Types

For organisms to survive they must sort themselves into favored environments, or habitats, in which they can live and prosper. Individual organisms have relatively few choices in this matter, they must either find a suitable habitat, adapt to an unsuitable one, or die. Few organisms can adapt to completely unsuitable environments in any appreciable way except for humans, who do so by using fossil fuels to live in incomprehensibly hostile environments such as the Antarctic or New York City.

Insects that fly or move freely about have some ability to locate new and potentially hospitable habitats. Although most insects are limited in distribution by their size and flying abilities, they receive supplemental traveling help in several natural and unnatural ways. Winds blow insects, sometimes for great distances. Birds may carry insects from place to place by eating fruits and berries with insect-infested seeds. These seeds are passed undigested from the bird, and the insect emerges perfectly normally after its coach-class trip. Similarly, we gardeners occasionally distribute insects in our seed exchanges, although, hopefully, we do so only in envelopes. Gypsy moth pupae, true to their name, are often found attached to vehicles that have traveled hundreds or thousands of miles. Consider the legalized, regulated worldwide trade in fruits, vegetables, seeds, cut flowers, wreaths, bulbs, plants, crates, boxes, baskets, and even furniture, all of which can be infested with insects; then consider the illegal, unregulated trade in these commodities. Insects get around, whether on their own or with the help of birds or humans. They do get around.

Insects do not always end up in a suitable habitat to survive. However, given the great numbers of insect species, the great numbers of insect individuals, and their ability to hitch a ride on or in almost any moving thing, insects have substantial odds in their favor to eventually get someplace they never were before. This applies directly to our gardens in both negative and positive ways. Positively, it means that if a garden offers the right habitat, say for butterflies or bees, then eventually, somehow, butterflies and bees will make it to that garden. Janice Emily Bowers, in her book *A Full Life in a Small Place* (1993), tells of desert creatures from far and wide reaching her city-locked, Tucson garden when offered a diversity of blooming plants. Negatively, insects unnatural to our gardens, such as the woolly adelgid, may arrive from exotic shores and devastate our hemlocks because there are no native, natural enemies that have coevolved to attack them here.

Insect species live in one of three basic habitats: terrestrial (soil level and above), subterranean (below soil level), and aquatic. Within each of these habitats, there are all sorts of subdivisions, or micro-habitats, and interfaces between them. Additionally, insects may have different requirements at different points in their life cycles—dragonflies, for example, live underwater as nymphs and in the air as adults. Therefore, many insects must be able to find several different suitable habitats during their lifetimes and then must be able to find their proper niches within that habitat. A niche is the sum of all life-sustaining interactions required for an organism: physical space, temporal space, food, competition, and the like. These niches can be extremely specific for some species, but for others they can be more flexible. Some parasitoids, for example, must find a particular host at a certain stage in a specific environment or they cannot repro-duce. Other parasitoids may be able to use completely different hosts in the same habitat and still produce offspring. A great deal of survival depends on the species of insect involved because some, even in the same genus, have greater tolerances for habitat diver-sity than others.

Terrestrial

Terrestrial habitats include the soil surface and above. Obviously, this is a fairly large area—nearly as large as all outdoors. Insects have been found from the Arctic Circle to the Antarctic, and there are re-ports of insects living at elevations up to 5030 meters (16,500 feet).

Honey bees have been reported flying at 10,670 meters (35,000 feet), but they were probably being propelled willy-nilly by air currents and were hopelessly lost as well. Still, it is fair to say that insects are widespread as adults and immature forms in the terrestrial world.

When an insect enters the subterranean or aquatic worlds (next sections), it is generally as a nymph or larva that will eventually reenter the terrestrial world as an adult. There are countless exceptions, of course, as noted below, but many insects that appear to be ground dwellers live primarily at the soil-humus-air interface, and many insects that are water dwellers must eventually enter the terrestrial world to mate and reproduce.

It is not necessary to dwell on insects in the terrestrial habitat because most of this book is about those kinds of insects. It should be stressed, however, that the terrestrial system can be divided into numerous categories and subcategories depending on the criteria being used. For example, using abiotic (that is, nonbiological) components such as temperature, moisture, climate, altitude, and latitude, the world can be carved into floral or faunal areas, much as gardeners have done for years with our hardiness zone maps and now our heat zone maps. We can also divide the world by its biotic (that is, biological) components along the lines of dominant species of plants and animals or associations of plants and animals. Of course, the biotic factors are dictated by the abiotic factors, but the end result is a number of intergraded associations of biotic and abiotic factors that biologists call realms, zones, regions, biomes, ecotones, and other names based, in part, on the scale at which the observer views the association being studied. Some such systems divide the Earth into vegetation types based on the predominant species of vegetation or the associations of species. In this system, a major category might be based on grasses and include savannas (dry grasslands with trees), prairies (dry grasslands without trees), and meadows (wet grasslands without trees). Alternatively, a desert habitat might be subdivided by plant associations such as oak-juniper or juniper-pine.

Whatever this breakdown—whether by plant or by habitat—it is almost certain that the insects present will be taking advantage of these habitats in one of two ways. Specialists are those insects that are restricted by some aspect of the habitat—whether host plant, host insect, temperature, or geography. Specialists will only be found living naturally within a particular prescribed set of environmental limits. To be extreme, there are insects that can live only near glaciers, prey on only one or two species of aphids, or pollinate only one species of plant. These specialists are limited in distribution by some aspect of their environment (taken in the broad sense), and if they are blown by winds or fly into unfamiliar territory, they will be in serious trouble. Generalists are insects that are not restricted by habitat or food supply. These insects may be found living naturally in many places. For example, a predator that feeds on many different kinds of aphids has little difficulty adapting to new situations when they arise.

The concept of adaptability to a habitat has many components to it and is not a straightforward notion at all. Insects often fall between categories of specialist and generalist. If it is blown onto a snowfield, the generalist aphid predator will die as surely as the glacier insect will if it is exposed to heat. An insect may be specialized in diet, but be adaptable to temperature—or the reverse. The point of highlighting these possibilities is to emphasize the numbers of niches within any given habitat and within any given environment. The niches for terrestrial insects are endless. When these possible niches are augmented by various life stages diving underwater or underground, the effect is multiplied almost beyond imagination.

Subterranean

Many insects live below soil level at least part of their lives, but few spend their entire lives underground. It is common for some insects —grasshoppers, for example—to lay their eggs in soil, but the hatching nymphs exit to the terrestrial world to live and feed on plants. Some insects use the soil as a nursery for their larvae, theoretically giving them a degree of protection from the harsh world above ground. Ground-nesting bees and wasps develop in cells on a food source collected above ground. These larvae develop on pollen (bees) or prey (wasps), pupate, and leave the ground never having been exposed to a single bite of subsoil sustenance. Antlion larvae live in underground pits and trap aboveground prey as they stumble into the slippery-sided pit. Similarly, tiger beetle larvae live in holes from which they grab insects that pass by.

Some insects lead extensive subterranean lives, but still forage daily for food in the terrestrial world. Ants are a common example. Some ants even forage for leaves above ground, then turn them into an underground fungus farm. Conversely, thousands of insects live above ground as adults, but their larvae are completely subterranean. Perhaps flies and beetles offer the most diverse temporary soil inhabitants in the form of larvae. These insects spend their immature lives feeding on decaying materials, plant roots, and other insects.

It is extremely rare to find insects that never leave the soil. Termites, most species of which basically spend their entire lives underground (even, ironically, when they build twenty-foot mounds above ground), exit the soil for short periods to mate and initiate new colonies. Cicadas, some of which spend precisely thirteen or seventeen years underground, emerge into the daylight to reproduce. Some aphids, scale insects, and a beetle or two are believed to live through continuous cycles underground, but for the most part these sorts of insects are rare. To be certain, insects use the subterranean world in great numbers, but it is largely as a staging area for one or more parts of their lives, and they emerge into the terrestrial habitat when necessary.

Aquatic

As with subterranean insects, few water insects spend their entire lives submerged beneath the surface. Most, such as mosquitoes, midges, blackflies, caddisflies, stoneflies, mayflies, and dragonflies, are aquatic only during their immature stages and exit to the land to mate and reproduce. Even adult beetles that spend both their immature and adult lives in the water can fly away to a new pond or lake if they have a mind to.

As with terrestrial insects, those that live in an aquatic environment have endless sorts of habitats. First there are the basic structural components of a water feature itself: its surface layer, the body of water beneath the surface, and the substrate upon which the water sits. Each of these subdivisions can be found in seamless varieties within rivers, creeks, stagnant ponds, freshwater and saltwater lakes, springs, and even geothermal waters. Insects have been reported living in hot springs at temperatures up to 60°C (140°F) and they have been collected at the bottom of Lake Baikal in Russia at a depth of 1360 meters (4462 feet). Not surprisingly, this habitat diversity is capitalized upon by insect species, as both adult and immature forms in a number of orders. Most surprising, perhaps, is the fact that the vast oceans of the world have not been invaded by more than a handful of insects.

Whatever insects have to do to survive in today's world, they have been doing for longer than anyone can remember. Fossil evidence records primitive Collembola (springtails), an order that is still living, from 400 million years ago. Those ancient standbys the cockroaches—at 300 million years old—are mere babes in arms compared to the springtails. Living insect species recognized today are known from 15,000- to 20,000-year-old fossils. Before flying reptiles and birds even thought about soaring, the largest flying creatures in existence were insects.

Insects, it seems, have staying power—they plan on surviving wherever they can. And one thing seems certain: a few million (or even billion) tons of toxic chemicals are not about to change their tiny collective minds. In fact, insects are evolving resistance to chemicals just about as fast as we humans can devise new ones. Perhaps the greatest of all celestial ironies is being played out. As we increasingly risk our own health in a war of toxic terror—a process of unnatural selection—we are encouraging insects to become ever more dominant life-forms.

A perfectly camouflaged female ambush bug sits atop goldenrod to await dinner, which she will grab with her front legs. As a male approaches cautiously from beneath, his tiny mind is overcome with two opposing desires: becoming one with the female at one end, while staying out of range of the other. Predators will be predators, after all, and a fine sense of balance must exist between being a mate and ending up as dinner.

PART II

The Ecology of Gardening

> We have got things all twisted around. We try to manage the
> whole planet when our own backyard is a mess.
>
> EVAN EISENBERG
> *The Ecology of Eden,* 1998

Ecology is the study of how biological organisms interact among
themselves and with their environment. Whereas physics, chem-
istry, astronomy, and mathematics may provide answers to mind-
boggling questions based on physical properties and laws of the universe, ecology works in the biological realm—where discovering
universal truths has proven a difficult undertaking, even in the most
basic instances.

For example, if someone asks a biologist the simple question "Do
bees sting?" he falls all over himself trying to give an answer that
any five-year-old child could produce in an instant: "Yes, bees
sting." But biologists know that, technically speaking, this answer is
wrong, or at least partially wrong, or at least partially not right. Biologists tend to qualify their answers to cover every possible degree
of correctness. In so doing, we sometimes end up confusing everyone, including ourselves. With regard to whether bees sting, for example, to be technically correct we must answer, "Only female bees

sting because males do not have a stinger." However, many bees are so small that, when they do sting, a person does not feel even the tiniest barb. Thus, we are left with a twist on the age-old question "If a bee stings, but we do not feel it, does it still sting?" Then again, not all female bees can sting, so we biologists must qualify our answer further and say, "Only some female bees sting, whereas some others do not." There is a whole group, the stingless bees, that does not sting but bites. Many otherwise rational folks might fail to detect the fine distinction between being stung (by the butt-end of a bee) and being bitten (by the head-end of a bee). But, to be biologically correct, all of these elements should be included when you answer the question "Do bees sting?" Such precise imprecision is the curse of being a biologist.

When we ask ecological questions, one of the basic tenets ought to be about how organisms interact in the natural world. Because humans tend to see themselves as the center of the world, however, ecology sometimes takes on a meaning such as "the study of the balance between our needs and the needs of other animals and plants" (Mounds and Brooks 1995). In truth, the study of ecology generally has nothing to do with human needs and everything to do with nonhuman interactions. Realistically, however, we humans view many things from what might be charitably called a "needy perspective." Thus, we often perceive ecology as a tool to solve the dilemma "Now that we've screwed up the environment, . . ." Our approach often is reactive rather than reaffirming, as it should be.

Many gardeners forget that, although the garden is a piece of earth that has been artificially (and only temporarily) tamed, it is still subject to the laws, or whims, of nature. Unless eternal vigilance is practiced, any garden will quickly revert to a tangled mess (or a truly natural garden, if you prefer) in less time than it takes to fire up a weed-whacker. When given a choice, most gardeners would be far better off learning a few simple facts of nature than spraying three quarts of noxious chemicals into the areas in which their children and pets play. In part II, I present these facts of nature, with the profound desire that more gardeners will attempt to envision their gardens as ecosystems of organisms that commingle with profound and positive consequences.

If we gardeners can understand the complicated, living structure that insects bring to the garden as a whole, then we should have a better perspective from which to judge these mostly maligned creatures. Insects are an integral part of how a garden ought to function, if we would only let it. Insects are a part of the balance, and thus the stability, that characterizes a garden of repose—as opposed to a battlefield of insecticidal woes. To better integrate the garden into the real world we live in, the gardener needs to understand and, more importantly, to accept the stability provided by the complex interactions of insects and plants.

In part I, I presented an overview of what insects are, how they are classified, and how they grow and reproduce. In chapter 4, I discussed their various types of feeding behaviors and the basic categories of habitats in which insects live. Hopefully, this general knowledge of entomology will provide insights into what insects do in our gardens and why we might better appreciate their reasons for living in a place we have constructed essentially, we believe, for ourselves.

In chapter 5 of this section, I discuss the function of insects in the garden, that is, why they are important and why we should encourage more insects to move into our gardens. In chapter 6, I give a practical overview of the categories of interactions between insects and plants that we may find in our gardens. This is a primer of real-life insect-plant relationships. In chapter 7, I discuss interactions between insects. In chapter 8, I explore the way that humans perceive insects and some ecological principles that guide the way insects interact with each other and the garden. Finally, in chapter 9, I develop the concept of plant and habitat diversity as it relates to insects and the ultimate stability, or balance, of the garden.

In my view, one of the basic goals of a gardener should be to achieve a healthy biological balance within the garden. Although this is not a difficult undertaking, it requires two essential ingredients that a gardener ought to have in great abundance. One essential is time. If a gardener can plant a tree with the knowledge that it will take ten, fifteen, or even fifty years to reach maturity, then the same gardener should be willing to invest a few years in establishing a balanced garden. The second essential is patience. It does no good if the gardener is willing to wait ten years for a tree to mature if he goes out every third week and chemically exterminates every living being in the garden simply because there are holes in the edges of six leaves. As we shall see, the intelligence of patience tempered by time far outweighs the foolishness of the quick-fix remedy.

5

The Function of Insects in the Garden

Let's assume that you are an insect living in a garden. What would you be doing there? Chances are, it would be one of the same three basic life-sustaining activities you would be doing if you weren't an insect living in a garden: eating, seeking shelter, and mating. (The exact sequence of these functions I leave entirely up to your imagination—either as a human or as an insect.) Of course, we humans believe we are above the basic facts of life, but take the following tests and judge for yourself. First, try going without food for a day or two, which will convince you that much of your life is devoted to pampering your stomach. Second, try going without shelter for a month or two, which will show you how artistically creative you might feel. Third, try going without sex for a generation or two and see what happens to your family tree (or, more accurately, watch your family tree fall over). Without food, shelter, and sex the human race would die out in a geological moment. So it is with insects and with all animals and plants as well.

Most of the time when we see insects, we convince ourselves that we are looking at nature's inconveniences—minor irritations of no particular significance. However, when we see elephants on television, in the zoo, or in our backyards (should we be living in Africa or India), we are awestruck by their enormous size, odd appearance, skin, feet, tails, ears, trunks, and, not without a little jealously, their ability to squirt water through their noses. Humans have been delighted by the antics of wild, exotic animals since the beginning of time, with only one proviso—that we do not have to suffer their brutish instincts. No one wants to be stomped on by an elephant, unless, like the circus performer, it's part of your job. So it is with lions, bears, tigers, alligators, sharks, and so on throughout the list of worldly creatures: we wish to enjoy them, but from a respectable distance.

Through movies, television, and mass-produced picture books, our last few generations have been exposed to animals found throughout the terrestrial and aquatic worlds. We have been conditioned to believe that all animals are wonderful—as long as they have fur, fins, or feathers or are cuddly, cute, or beautiful. It might be possible to say that everyone likes animals in their own special way. Some spend their lives studying animals, some like to eat them, some like to shoot them, some like to keep them as pets, and some like to breed them.

Why, you might ask, am I whining about elephants and feathers and fins, when this is supposed to be a book about insects? But that is just my point: insects *are* animals. The tiniest insect alive (two of which would fit on the period at the end of this sentence) has a head, wings, six working legs, antennae, a heart, a brain (of sorts), and reproductive organs all of which function to make this mere speck of life function as completely as you or I. We tend to forget these facts when we compare insects to the "real" animals with which we share the planet. For example, you never see People for the Ethical Treatment of Animals protesting the mass slaughter caused by mosquito-abatement programs. Nor does anyone protest the insidious and worthless electric bug killers that fry countless millions of innocent lives in our backyards—these devices kill scarcely a single mosquito, by the way, the purpose for which they are sold. No, we humans have almost no sympathy for any form of life that does not receive good press on television, and with the single possible exception of butterflies, insects generally do not. We hear about killer bees, invading fire ants, car-encrusting love-bugs, forest-killing gypsy moths, Japanese beetles, fruit flies, Colorado potato beetles.

So I return to my original premise: we are going to enter the world of the insects and see what is going on at gnat's-eye level. It is the only way I can convince you that insects live in your garden for a purpose and the purpose is equally as interesting and entertaining as looking at the relatively colorful buttocks of a baboon or watching a peacock strut his stuff.

Food Chain

By now you might be entertaining the notion that you have placed yourself in the hands of an eccentric madman who believes that insects have a beneficial significance far beyond that which they deserve. In this matter, I defer to a quote by the double Pulitzer Prize winning author and Harvard entomologist, Edward O. Wilson, who stated that "insects are the little things that run the world." Now there's a news flash for you. What happened to humans? Don't we run the world? To answer these questions would take philosophers and theologians centuries of hotly contested debate, but I can answer the question in one word: No. We certainly might ruin the world, but we do not run it, not by a long shot. If you would like to test my hypothesis, simply answer the following two questions: Can the world function without humans? Can humans function without the world? The answers are so obvious that I will not bore you with their explanations.

To prove that I can be a reasonable person, I will say that one of the many ways in which insects are valuable to the garden is by sacrificing their lives so that other creatures might live—a truly noble deed, if it were done consciously, which, of course, it is not. Insects provide the nourishment for a great many other animals, birds, for example. Many of us rather enjoy birds in our garden, they enliven our souls with their sonorous singing, intricate and daring flights, · painterly colors, darting shapes, the architectural delight of their

Mosquito larvae and pupae float in their aquatic environment. Both life stages are part of an interconnected food chain: adults are eaten by bats, birds, and predatory insects; adults and larvae are eaten by pond fish and aquatic insects.

nests (if we are lucky enough to have resident birds), and the bustling liveliness of their fledglings.

To be certain, many birds eat seeds, but just as many supplement their seeds with insects or eat insects almost entirely. Imagine a garden without robins, swallows, warblers, meadowlarks, chickadees, vireos, gnatcatchers, wrens, nuthatches, titmice, swifts, flycatchers, woodpeckers, flickers, cuckoos, waxwings, blackbirds, or orioles. Even hummingbirds and owls eat insects. Given a good bird population, as found in my yard, the gardener also ends up with a few birds of prey as well. A hawk near the garden always calls for a work break. On several occasions, I have seen pileated woodpeckers in my garden. That sighting called for a week's vacation!

Another group of animals most welcome into anyone's garden is frogs, toads, and salamanders, all insect-eating amphibians. My garden has at least half a dozen species of resident frogs and toads (no salamanders yet). Insects form a major part of the diet of these animals, and I have numerous containers of water and a couple of small ponds that help attract them. All these creatures arrived on their own; I did not bring a single tadpole or egg into the garden. I do not know which was more influential in attracting these amphibians to the garden, water or insects, but certainly having both makes them want to stay.

Water also provides opportunities for organisms associated with insects. In an established garden pond, you will never have to feed the fish because they happily devour all the insects they can find, both under the water and at the surface. They will eat all the mosquito larvae and adults without hesitation. You may also be inviting the great blue or small green heron to come and dine on the fish in your pond. By feeding the fish, insects make such spectacles possible.

Many of us might also welcome lizards and skinks into our gardens, a group of insect-eating reptiles that has not yet appeared in my eastern garden. As a child growing up in the western United States, it was not uncommon to have fence swifts and alligator lizards pop up in the garden from time to time. These were always curious creatures, and any right-thinking young boy would dearly have loved to find such wondrous beasts in the garden.

Now, to nearly complete the big animal elements of the food chain, we come to an insect-eating creature that I hesitate to mention, for as surely as I do, the less enlightened of you will throw this book down and run off screaming. Please either hear me out or skip over the next few paragraphs. If it were not for incredibly bad press, this animal would be almost as cuddly as a rabbit, and it is a damn sight better for your garden. I am speaking of the unfairly disparaged and unwisely defamed bat. If you want a bug killer that actually works and costs nothing to operate in the bargain, then what you really want is bats. This does not mean that you must have bats in your belfry, but if you see a few flying about at night there is no need to get out your trusty flame-thrower and incinerate them.

Yes, bats are somewhat prone to rabies, but less so than raccoons and skunks, which no one seems inclined to kill at first sight. It is only because of disingenuous vampire movies and the fear of bats tangled in one's hair that rabies is used as the scapegoat for our primitive and superstitious fears. It gives us good reason to kill something of which we are basically afraid.

So let's set the record straight. In the United States and Canada, almost all bats are insectivores, that is, they eat insects. Those that are not insectivores are nectarvores—they sip nectar and pollinate our plants. Can that be so scary? Wouldn't you pay big money to have someone come in during your evening barbecue and quietly consume 500 mosquitoes per hour, which is what bats can do? Consider yourself privileged to have a free air-clearing service. Besides, much of this consumption is occurring without the gardener's knowledge. You rarely see bats unless you look for them. They are circling in the heavens over your backyard, making life a lot better for you. As for the fear of rabies, simply do what you would do if any wild animal came too close: move away from it. It is the animals with rabies that do not have enough sense to run (or fly) away from humans, as they normally would.

In addition to the bat, the only insectivorous animal not generally welcomed into the garden is the mole, of which I've had an outbreak or two. In their quest for insects, moles do not purposefully destroy our plants by dislodging them or sheering roots, nor do they turn the garden's surface into chaos with their underground swimming. Still, they are a distraction, although I've never had any significant problem with moles that time did not correct. Because they prefer moist, humus-filled, friable soil, moles tend to leave the garden as soon as it returns to its normal, uncultivated, neglected state. In spite of minor troubles, moles eat a lot of white grubs and in so doing reduce the larval populations of root-feeding beetles.

Gardeners should note that the commonly destructive large animals found in the garden, essentially all pests, do not routinely eat insects: gophers, squirrels, chipmunks, deer, rabbits, mice, and woodchucks. Put in perspective, if a gardener might be said to have a choice, I believe that the most sane, right-thinking sort would prefer to have insect-eating birds in preference to rabbits; insect-eating frogs in preference to gophers; insect-eating fish in preference to woodchucks; and, yes, even insect-eating bats in preference to deer.

I have purposefully left out the major members of the insect-eating clan because they are the subject of this book. That is, the most common food chain interactions you would find in the garden are insects (and spiders) eating other insects. I leave them out here for two reasons: because I devote chapter 7 specifically to a study of their interactions and I want to accustom you to thinking about insects as real animals. I am never quite certain that people believe entomologists when we say nice things about insects. Hopefully, by the time I do, you will almost believe me. If you are one of the intelligent sorts of gardeners—and you must be if you are still reading this—then I hope you have seen that insects are integral parts of a complicated food chain. The garden is simply a small slice of a vast world of interactions. To have insects in the garden is to have life in its many forms. And what is a garden if not a manifesto of life?

Unless . . .

Decomposition

Unless it is also a manifesto of death. All living things will inevitably die. (I did not make the world, so please do not shoot the messenger.) Death and dying are natural processes of life, as we have seen in the preceding section. In the food chain, it is the fate of insects that a great many of them do the dying—and in ways that could be considered unnatural. In nature, of course, being eaten alive is natural, but so is keeling over as a result of the sheer exhaustion of living. Not all animals end up on the dining room table.

In a real garden, plants and animals are always dying, either in total or in bits and pieces via the process of senescence. Little bits of roots, leaves, stems, twigs, and branches are continually sloughing off in a process of renewal. Sometimes whole plants die. Annuals die at the end of the season. Trees (even redwoods) die. Herbaceous plants and bulbs die back to the earth each year. Animals die of old age. Insects, spiders, worms, snakes, millipedes, frogs, birds, even rabbits die. They all reach a point at which they simply fall over and do not get back up again. With the exception of what we call yard debris—the excess by-products of our plants—we gardeners never view the evidence of what might be termed the "remains of the day." I do not know any gardeners who want to look out into their back garden and see a pile of dead things lying about for years and years.

Nor would they wish to see the other staple of the vicissitudes of life, large piles of excrement, unless of course, it is horse or cow manure. The by-products of living, in addition to many dead bits and pieces, are a lot of what we entomologists delicately call frass. This by-product goes by other names, but feces, droppings, manure, and doo are the ones used by polite gardeners. In some societies the use of human waste direct from the source is common, whereas in others it is sanitized and sold as soil amendment. Very few gardeners devote endless hours pondering what happens to the wastes of all the animals in and around their garden or the dead bodies that ought to be piling up there in boundless heaps. That is where I come in.

The time has come to talk about decomposition. Of course, decomposition is not particularly polite conversation at the dinner table unless one approaches it in terms of, say, the compost pile. Here we gardeners love to talk about decomposition. The more decay and the faster it happens, the more we love it. (This is one topic, by the way, that separates gardeners from normal people. Try it at your next gourmet supper; you'll spot the gardeners in an instant.)

When I talk about decomposition, however, I am not speaking of the compost pile. That's too obvious. I am speaking of all the once-living detritus and waste products that rain down on your soil (and arise in your soil) minute by minute, day by day. The entomology textbooks point out that if all the progeny of a pair of fruit flies survived and all reproduced normally for one year, there would be a ball of fruit flies 96 million miles in diameter, give or take a foot. (I am not making this up, and, although personally I find it hard to believe, I am not about to do the math myself.) Oddly, these textbooks never calculate the waste products, which might add another couple million miles to the ball of flies. Anyway, my point is that the primary purpose of life is to recreate itself. It does so quite excessively (and messily) at times, yet we rarely see the end products of life, namely, effluents and bodies, piled up in our gardens. Why is that?

The reason, in large part, is insects, along with bacteria, fungi, and worms, which perform a welcome service by helping break down the bits and pieces of life that we humans do not particularly relish finding in our gardens. The list of such things is endless, but here are a few examples of the bodies that could be found dead in any garden: squirrels, rabbits, mice, chipmunks, rats, moles, voles, gophers, woodchucks, armadillos, raccoons, opossums, cats, dogs, birds, snakes, frogs, toads, lizards, spiders, millipedes, centipedes, and insects. (The precise make-up of any garden depends in large part on its geographic location.)

The first line of approach to the decomposition of all these dead bodies is the insect. So predictable, in fact, is the arrival and departure of insect decomposers that an entire branch of entomology has been developed to use this chronology for solving homicides: forensic entomology. In exacting and temporal detail, forensic entomologists know that the succession of insect decomposers is about as follows: first come bluebottle flies and house flies; then flesh flies and a few predatory wasps (such as yellow jackets and hornets); then hide beetles, flies called cheese skippers, and moth larvae; these are followed by silphid and hister beetles; then some different sorts of flies; then more beetles and clothes moths; and, finally, the general scavengers, including more different sorts of beetles, wipe the slate clean (even if a few bones might remain).

I will not delve into the finer points of body disposal, but it is an important and natural function for which insects and microbes provide service without whining. There is one heck of an assortment of insects tidying up your garden at no cost to you, and 99 percent of

A carrion beetle chews on a mouse carcass. The toils of some insects help keep our gardens tidy. Were it not for these insects, our gardens might be knee-deep in dead bodies and waste products.

the time you don't even know it. Be thankful and consider the alternatives.

Another thing to consider is dung, which is squarely up there with dead bodies. Dung-eating insects are called coprophages and are not the subject of great discussions in polite circles. In the vast scheme of things, coprophages are very important insects because they clear the natural landscape of what otherwise few people—gardeners excepted—want. In our gardens, they basically serve to break down the unwanted by-products of our pets and the odd fox, raccoon, and bear that wander through the property.

There is not much nutrition in waste products (in case you should be curious) and they dry out quickly, so insects that depend on such things are exceptionally gifted at detecting their meals as soon as they drop. This is accomplished by a combination of fragrant chemicals given off by the entree (shall we call it) and chemical receptors on the antennae of the insect. The two main groups of insects that feed thusly are the scarab beetles (sometimes called dung beetles, for obvious reasons) and some groups of flies, such as the house fly, latrine fly, dung fly, and eye gnat—don't dwell too long on that last one. Most commonly, the larvae of beetles and flies live in dung, but some adult scarabs form feeding balls from which they apparently strain microbes dining in the manure rather than any particulate matter in the ball itself. Fly larvae feed on droppings where they drop, but some scarabs are famous for a dung-ball-rolling ballet in which they form a ball from freshly fallen feces, roll it to a spot, dig a nesting chamber, put the ball in it, lay an egg on or in the ball, then backfill the tunnel. Some less captivating scarabs simply burrow under the dung, dig tunnels, and backfill them with packed manure upon which their larvae will hungrily fatten up. As with much of human existence, perhaps some life is not so much concerned with daily chores as with the panache of the thing.

So now we've removed the dead bodies and their once lively by-products. What's left? Little bits and pieces of things, that's what. Here is where the scavengers, or detritivores, come in. Some insects specialize in eating dead plant parts. Humivores, including some

Blow fly larvae eat animal dung. Some might call this service beneath the call of duty, however, nutrition often remains in what others leave behind.

termites, eat humus in the soil and depend on microbial symbionts (that is, different organisms that intimately live together for the mutual benefit of both) in their guts to process the organic materials and dead arthropod and fungal bodies they find. Termites may be vilified for eating your house, fence, or furniture, but in the garden they help recycle all manner of organic plant materials back into the soil, including general plant litter such as leaves and twigs and even some animal wastes. The nutrients locked up in these dead

Head over heals, this scarab beetle dutifully rolls a ball of dung on which its larva will eventually feed. Sometimes we humans take such selfless acts for granted. Where would we be if no one cleaned up after us?

bits and pieces are released into the soil to be recycled again and again through living plants.

Along with termites, some beetle larvae, fly larvae, and carpenter ants feed in dead tree trunks and limbs. These insects help to more quickly break down the apparently solid lumps into elemental components. Many of these larvae do not actually eat sound wood or even dead, dry wood, but rather consume fungi that grow in already decaying wood. Therefore, before you blame an insect for killing your trees, you must always ask which came first, the fungi or the fly?

Pollination

Let us now shift to something a little more pleasant, the subject of pollination. We all think we know what pollination is, but do we really? We might vaguely be aware of the technical process of pollination: the transfer of male sexual cells (pollen) from one flower to the female sexual cells (ovules) of the same or another flower. But if you believe this is the essence of pollination, then you are equally likely to believe that milk is squeezed from cartons, apples grow in plastic bags, and peas come frozen in boxes from the North Pole. It is far too easy to dissociate the practical act of pollination from the metaphysical act of being, because without pollination you and I would not be here.

It is not simply that pollination is one way—one very important way—for plants to reproduce themselves. Without seeds, after all, there would be considerably fewer plants around. There are other reproductive methods, based on vegetative growth, for instance, but when it comes to examples of how plants survive on such reproduction, they are not encouraging. It could be argued that a fine fruit tree is propagated vegetatively. But we do not eat fruit trees, after all, we eat their fruit, and pollination is required for the formation of fruit. It could also be argued that potatoes are propagated vegetatively, but they are first developed by hybridization, which requires pollination. Pollination is needed to engineer the natural characteristics of fruit trees, vegetables, and grasses so that they bear over longer periods of time; in different geographic regions; or larger, more nutritious fruits or grains. In fact, it is difficult to imagine any food source that does not need pollination at some stage merely to develop a better or more productive type that can then be propagated vegetatively or by seed. Whether we think of pollination routinely carried out on a natural and daily basis or pollination used by humans to develop new or improved varieties of plants, the act of pollination is a basic process that keeps us alive.

Whereas many of us might intuitively realize the importance of pollination, we seldom know any of its finer details, delight in the process, or acknowledge its significance to our well-being and our very survival. But we should. According to Stephen Buchmann and Gary Paul Nabhan (1997), who unashamedly rhapsodize pollination in their entrancing book *The Forgotten Pollinators,* one in every three mouthfuls you swallow is prepared from plants pollinated by animals. Most, but not all, of those animals are insects. Think about that for a while, and you may begin to appreciate insects, even if for the utterly selfish reason that we would be much worse off should they disappear. And according to Buchmann and Nabhan, pollinators are disappearing at an alarming rate.

Before I address the subject of pollinators, their disappearance, and why we should encourage these insects for our own good, I need to clear up a few misconceptions. The first has to do with honey bees (about the only pollinator that comes to mind when you speak to some folks), and the second has to do with practically all the remaining insects.

Buchmann and Nabhan argue that severe reductions in honey bee populations—up to 50 percent—are occurring in the United States due to attacks by tracheal mites that infest their breathing system (the trachea) and varroa mites that attack their bodies externally. Honey bees suffer heavily from fungal, protozoan, and bacterial diseases as well. Africanized honey bees have discouraged beekeeping in the southern parts of the Southwest and may continue to do so if they move farther north and east. Not only are Africanized bees a bit tougher to manage, just imagine the potential lawsuits to which a

A long-horned beetle feeds on meadowsweet. This beetle may also be affecting pollination in an undirected, or accidental, way. Insects often pollinate flowers while they feed on nectar and pollen.

beekeeper might be exposed. Honey bees are important, of this there is no doubt. For anyone who wants a fair, although somewhat biased, view of the subject, I suggest reading Sue Hubbell's books, *Broadsides from the Other Orders* and *A Book of Bees*, both of which should be countered with Buchmann and Nabhan's *The Forgotten Pollinators*.

Honey bees, however, represent but a fraction of the number of insects that pollinate plants, and in many cases honey bees cannot even get the job done. The world would not end if all honey bees were suddenly eliminated. According to Buchmann and Nabhan, there are about 80,000 other species of bees, wasps, and ants; maybe 20,000 species of butterflies and moths; 15,000 species of flies; 210,000 species of beetles; and a few thousand more animals including birds and bats that contribute to the pollination of approximately 240,000 species of flowering plants throughout the world.

With the publication of Buchmann and Nabhan's book arose an organization, The Forgotten Pollinator Campaign, whose mission is to inform and convince ordinary folks to consider the plight of the native pollinators of all persuasions, whether birds, bees, bats, beetles, or butterflies. The campaign's message is that through humans' ever-increasing pressure, we are degrading native pollinator habitats either through the use of pesticides, development for human habitation, or the conversion of land to cropland. Such changes in habitat can have a domino effect and lead to a succession of linked extinctions, wherein, for example, the loss of a pollinator could easily cause the loss of a plant species. These extinctions can also result in the loss of pollinators for our food sources.

According to Buchmann and Nabhan, the great and noble service that gardeners can provide for all pollinating insects is to give them asylum—to make the garden a sanctuary for some of the insects that help run the world. Immediately as I write these words, I hear the unbridled hubris of the typical homeowner and even the typical gardener, "What! Is he completely nuts? I will not have my garden full of those damn honey bees." And I agree with you. I don't want your garden full of those damn honey bees either. I don't even like honey bees. For all of you who like life's little ironies, I pass along this tidbit of information: the highly touted honey bee is not even a natural element of the Western Hemisphere. Every honey bee in every garden in every country in North, Central, and South America is the product of colonialist introductions from Europe, and later, as you undoubtedly know, from Africa. For all the gardeners who lean toward the native or the natural and wish to garden naturally, every last honey bee must be destroyed. They simply are not native.

There are two reasons for inviting pollinators into the garden, one of which is selfish and the other of which is unselfish. Naturally, I will discuss the selfish part first. If you are food gardener, pollinating insects increase your intake of vegetables and fruits as well as seeds for the heirloom vegetables. The gardener gains seeds for the next crop, nutrition in the form of produce, and perhaps prizes at the county fair, while the insect gains nutrition, shelter, and a place to raise its family. This is what biologists call a mutualistic interaction—both parties gain. The sorts of insects that will most effectively pollinate your plants are bees, wasps, beetles, butterflies, moths, and to a lesser extent flies.

Those of us who are not food gardeners, and I count myself here simply for lack of room to grow everything, appear to have few reasons to invite pollinators into our gardens. After all, flowers have a fixed size, they do not grow larger by being pollinated. Bulbs, shrubs, and trees are usually not grown from seed. Many gardeners plant store-bought seeds of hybrid annuals, from which seeds do not produce identical progeny and so are not often collected. But if we are seed savers, say of heirloom varieties or old-fashioned and cottage garden flowers, the act of pollination increases seed set and we profit. On the whole, however, perhaps most of us rely more heavily on the seed and plant catalogs and our local nurseries for an endless supply of garden needs; rarely do we need garden-pollinated seeds to propagate our gardens. We are the ones, then, who need to be unselfish in our approach to our pollinator friends. We are the ones who can provide sanctuary to the ever more frequently homeless

A bumble bee alights on a nasturtium, with its neat packet of pollen collected from previous flowers scraped into a ball on its hindleg.

After considerable work, the bumble
bee is covered from head to tail with
copious amounts of pollen, some of
which will be carried to the next flower
and effect pollination. Some will be
added to the pollen packet and will go
home to the nest to feed the larvae.

pollinators, even though we appear to gain no immediate benefit.

From the perspective of The Forgotten Pollinator Campaign, it is the gardeners of the world who can open their gardens to the pollinator refugees, who can provide temporary or permanent shelter until humans refine our outlook on the natural world. By actively sheltering pollinators, we gardeners remind ourselves that we have the power to positively overcome some of humankind's more destructive tendencies. Additionally, our gardens provide a teaching laboratory for young children to connect with an ever-vanishing natural environment. Our gardens might provide a network of urban and suburban biological corridors that link more protected sites and allow pollinators to move freely from one natural area via our gardens to other natural areas. And finally, all lofty, Earth-saving notions aside, you might wish to encourage pollinators in your gardens simply because they are more interesting than any television show you can imagine. (Aspects of attracting pollinators to the garden are discussed in chapter 11.)

Balance

The importance of balance in a garden is rarely, if ever, discussed in gardening books, unless one is evoking the artistic elements of design, which I decidedly am not. I am referring, instead, to the ecological balance that most gardens are designed not to have. Relatively few garden writers have dared suggest that gardens be planted to achieve an ecological balance. Oh, there are those that tout natural gardens, butterfly gardens, water gardens, xeric gardens, organic gardens, or companion-planting gardens; in fact, these gardens are becoming somewhat de rigueur. In reality, the concept of ecologically balanced gardens is relatively new and perhaps one of the best gardening ideas to arise from the latter half of the twentieth century.

I'm not certain where or when the idea arose that it might be desirable to work with nature instead of against it. This was certainly not the purpose of the great landscape parks of eighteenth- and nineteenth-century England, where nature was completely rearranged

Sometimes a flower's nectar is out of reach for bees with short tongues. This thieving bee simply cuts straight to the bottom of things by going through the side of the flower and collecting nectar without providing pollination in return.

from the ground up, so to speak, to suit the aesthetic eye. Somewhere along the way, the picturesque, or natural looking, became synonymous with nature. However, there was scarcely anything at all natural about these parks. They were simply an artistic rendering of what a romanticized nature might look like if it wanted to be a Turner painting. As a naturalist might say, "Where's the ecology?"

If one looks at almost any gardening book, the subject of insects is either neglected completely or couched in polite terms of total annihilation. (With the exception, of course, of the butterflies. We always have a place in our gardens for butterflies, unless they happen to eat our plants to the ground, then they, too, are evicted as helpless larvae from their gardens of Eden.) Trying to keep insects out of the garden, although a favorite topic of gardeners, is impossible even if we use enormous amounts of toxic chemicals coupled with barriers of some kind. And what's the point? As the preceding discussions demonstrate, insects provide a number of essential and useful free services to the garden as well as make our gardens and our lives a little more interesting. We should encourage, not discourage, insects to flock to our gardens. In fact, insects do this even without our encouragement, but we seldom bother to look.

One fellow who did bother to look was the entomologist Frank Lutz, who worked at the American Museum of Natural History in New York City. Within a year's period, Lutz recorded almost 1500 different species of insects in his suburban garden, and even he admitted this number did not represent the tiny, difficult-to-see sorts —the balancers of the insect world. Lutz's figure was not the number of individuals, which would have been hundreds or thousands of times that figure. Lutz's number of different species is, in fact, one measure of the diversity of insects that lived in his 75-by-200-foot garden. In a footnote to his compulsive insect counting, Lutz was quick to point out that these insects caused absolutely no harm to his garden. To prove this, he proffered three medals and a certificate of achievement in four years of garden contests sponsored by a New York newspaper. Not bad for a yard full of bugs.

As we shall see in chapter 9, the more balanced a garden is in terms of insect-insect interactions, the less the gardener will have to worry about the tribulations of any one insect. If your garden existed in a state of balance—perhaps grace would be a better term— you would be able to ignore insects almost completely and get on with the real purpose of your life, which is to weed.

6

The Interactions between Insects and Plants

When an insect awakens, the first thought that comes to its puny mind is not how much trouble it can create for you, the gardener. The concept of trouble is confined entirely to the mind of the gardener and his philosophical outlook on such things. As for the insect, all it wishes to do is eat, develop, and reproduce. In fact, it does not even know it wants to do this. An insect does what its genes instruct it to do and has no preconceived idea of what it wants to do. In metaphysical terms, an insect simply *is*—it is the living essence of *is*ness.

Insects play a diverse and complex role in our gardens, of which most gardeners have little knowledge. At most we might be familiar with the aphids that attack our roses or the butterflies we see fluttering by (if we are lucky); we might note a few ants in passing or a hornworm on our tomatoes. But essentially gardeners know very little about insects. Ignorance can certainly be blissful, but a good dose of reality now and again is more useful. Ignorance should be overcome—in some cases, at least—and I believe that our gardens will be better off for the effort.

It may seem superfluous to state that the focal point of all garden interactions is the plant—no plant, no garden. (Some could argue and point to a contemplative gravel garden, I suppose, but then they would not be reading a book such as this.) The horrible truth about plants is that plant-eating creatures, or herbivores, attack them.

Deer come to mind, as do elephants, giraffes, and bison. Most gardeners should be exceedingly grateful that the most abundant herbivore they will run across in their gardens is the simple-minded and exceptionally tiny insect. Things could be a great deal worse.

In this chapter, we will examine the real-world interactions of insects that eat plants. Such insects are phytophages and their behavior is phytophagy. Sometimes they are called herbivores or herbivorous insects. Occasionally, it is just simpler to say plant feeder or plant-feeding insects. In this chapter, I demonstrate the vast array of ways in which different insects use plants that nature (and the gardener) provides for them. In turn, this basic information will enable us to approach a slightly more challenging appreciation of how insects interact with each other (chapter 7) and how the basic diversity of plants contributes to ecological balance in general (chapter 9).

Herbivores

As discussed in chapter 4, insects have mouthparts designed for chewing (as in humans) or for piercing and sucking. Based on these basic types of mouthparts, the gardener might have an impression that insects ought to be limited in their feeding abilities. But insects know no boundaries when it comes to feeding on plants. This is because they have refined these two simple feeding methods into a virtuosity of dining possibilities—at least half of all insects are thought to be herbivorous.

The manner in which insects feed on plants may be seen from two opposing points of view—each of which is equally valid. One way to examine these plant interactions is to focus on the part of the plant being eaten, for example, the roots, leaves, fruits, or seeds. Plant eating is the primary way in which food gardeners and farmers—those who depend on producing a crop—view insects, and it is the most likely way gardeners would first take notice of an insect. This viewpoint keys in on the perceived damage done to some essential plant.

Another way to examine herbivorous behavior is from the viewpoint of the insects, which put their chewing and sap-sucking abili-

Some herbivorous insects have strict feeding preferences, particularly in one stage. The adult of this gulf-fritillary can survive on nectar from different plant sources, but its larvae can feed only on foliage of the passionvine.

ties to work in a variety of different ways, for example, by mining or rolling leaves, causing galls on different parts of a plant, or boring into stems or seedpods. In the following outline of insect-plant interactions, I combine both approaches, first selecting the part of the plant attacked and then describing the different types of feeders a gardener might find on that part of a plant.

We should keep in mind that, within reason, insects are relatively specific in their feeding habits. You will not commonly find root feeders eating leaves or leaf feeders boring into stems. But each insect feeding type (root feeder, for example) will include different species that may be restricted to different types of plant growth. For example, some species of root-feeding insects will attack tree or shrub roots, whereas others prefer perennials, vegetables, or grasses; some root feeders attack exotic houseplants or orchids, others feed on annuals. In truth, no simple approach to categorizing plant feeding can prove totally accurate, but the approach used below may provide a quick indication of what is likely to be happening in some parts of your garden at any given time. The purpose is to demonstrate the immense diversity of food options that plants offer to insects and to demonstrate the many different ways in which insects take advantage of that offer.

Subsoil Feeders: Roots, Tubers, Corms, Bulbs, and Rhizomes

Because much of a plant's structure is buried out of sight, we forget that it might be subject to insect feeders. The food gardener, perhaps necessarily more attuned to her environment, will certainly be on the lookout for cabbage maggot, carrot rust fly, or onion maggot, all fly larvae that feed on subsoil tissue. But the nonfood gardener might not be quite so savvy when it comes to underground insects that feed on various plant parts.

The classification of roots is delicately confusing, shall we say. But gardeners generally get the drift that there are a number of different types of subsoil structures, even if we do not, technically speaking, always know what they are. Few among us perhaps realize that dahlias, tuberous begonias, anemones, cyclamen, and sweet potatoes are considered root tubers (that is, an enlarged root), whereas true potatoes are stem tubers (that is, the enlarged tip of a rhizomaceous stem). Rhizomes are an underground horizontal stem, as found in bamboo, bearded iris, or ginger. Other modified stems are corms (short, swollen, underground stems), such as those found in crocus and gladiolus, or bulbs (short, flattened stems bearing fleshy, food-storage leaves), such as those found in onions, lilies, daffodils, and tulips. Not all of these divisions are as neatly delineated as one would want them to be, but the insects appear to be able to separate out the various parts to a fair degree of refinement.

ROOT FEEDERS—Perhaps the most notorious group of root chewers is the beetles. White grubs, for example, a collective term for the C-shaped larvae of scarab beetles, are frequently found in lawns where they chew right through grass roots. Well-known examples of these larval root feeders are the Japanese beetle, the June or May beetles, and the rose chafer. However, these are probably better known to gardeners in their adult forms, which chew to bits just about any part of an herbaceous shrub above ground. Other beetle larvae, such as the wireworms, are perhaps not so well known, but they feed on the roots of many different plants, especially in the seedling stage. Weevils, being among the most species-rich of all insect families, also contain many root-feeding species such as the Fuller rose weevil, strawberry root weevil, black vine weevil, and white-fringed beetle. Many of these are devastating root feeders. Flea beetle larvae, also called leaf-feeding beetles as adults, are not averse to feeding on the roots of corn, cucurbits, and grapes, among other things.

Aphids and their relatives the scales, mealybugs, and cicadas constitute another group of root feeders. Whereas beetle larvae are root chewers, the homopterous insects within the order Hemiptera are all sap suckers that drain liquid from the roots of their hosts. Additionally, almost all these root feeders, with the exception of cicadas, are found in loose-knit clusters or formless colonies. Included in this group would be root aphids such as the aster root aphid, which feeds on various asters as well as dahlias and cosmos; the corn root aphid, which—its name notwithstanding—feeds on oxeye daisy, chrysanthemum, sunflowers, and rhubarb; the pear root aphid, which attacks pear and quince roots; and the grass root aphid, which feeds on its namesake. There are many mealybug species that attack the roots of nearly all perennial shrubs, grasses, and some annuals. Many gardeners are familiar with adult cicadas simply from the male's heat-induced courting call. Although the adults are short lived, their immature stages (nymphs) live underground, sucking tree roots for anywhere from two to seventeen years, depending on the species. (For those gardeners interested in trivia, the cicada is the most reliably long-lived insect known. However, *The Guinness Book of Records* reported an adult wood-boring beetle that emerged from a timber in a house forty-seven years after the timber had been cut. Much of that time was probably spent in some kind of suspended growth pattern.)

The larvae of many kinds of flies also infest plant roots. The root maggot family, by virtue of its name, would seem to be one of the most obvious. Included in this family is the cabbage maggot, which is primarily a root feeder on cruciferous plants and which also bores into the basal stems of these plants and the swollen, fleshy parts of radishes and turnips. The carrot rust fly lives in the swollen roots of carrots and parsnips and the crowns of celery. The root gnats, contrary to their name, are primarily fungus feeders, but some species attack healthy roots. March flies also are primarily fungus feeders with a few root-feeding species. It is common to find numbers of fungus gnats associated with pots of dead and dying plants, for which the flies are blamed as root feeders. More likely the problem is the reverse: too much water, some root rot, and then fungus gnats feed on the decaying plant roots. (With insects, one must always be careful when placing blame—there is much that we do not know.)

GALL FORMERS—The formation of galls on plant roots is relatively uncommon, but it is known to occur. The most infamous example is found in an insect likely unknown to the general public. Members of this group are called phylloxerans, or sometimes chermids, and some species cause small swellings on grape roots. In the late nineteenth century, phylloxerans from American grape stock accidentally introduced into France had a disastrous effect on French viticulture, which was nearly destroyed. In addition to grapes, phylloxerans also attack the roots of willow, but given a choice between wine or willow, most gardeners don't seem to worry so much about the latter. Gall weevils cause swellings on the roots of various members of the brassica family, and gall wasps cause galls on oaks and rosaceous plants, including rose and blackberry. The largest group of gall formers, however, is the gall flies, the larvae of which attack such varied plants as orchids, irises, Virginia creeper, and even poison oak.

BULB, CORM, TUBER, AND RHIZOME FEEDERS—Several different insects feed on the swollen, underground parts of plants called bulbs, corms, tubers, or rhizomes, which are various modifications of roots or stems (as discussed above). The flies present a number of such feeders. The larvae of one species of crane fly are generalists and feed on many different kinds of bulbs, corms, and tubers. Adults of the bulb fly (or narcissus fly) and lesser bulb fly (the onion fly) would be mistaken for black and yellow furry bees or wasps. Their larvae feed on hyacinth and narcissus bulbs. Onion maggots, naturally enough, feed on onion bulbs. A few species of aphids invade the soil, including the tulip or iris root aphid, which feeds at the base of iris and tulip stems on the surfaces of bulbs and rhizomes and also on the roots. The woolly aphid feeds on gladiolus corms, as does the gladiolus thrips. This thrips first feeds on stored, dry gladiolus corms, then when the corm is planted in the ground the thrips migrate up the newly emerging stem to feed on aboveground parts of the gladiolus. The iris borer, a moth larva, bores into the rhizome (and leaf bases) of irises. Beetle larvae called wireworms feed on potato tubers, and larvae of the sweet potato weevil feed on sweet potato tubers.

All told, there are many different sorts of insects that feed on the

Two narcissus flies mate on a geranium flower. The larvae of these flies feed on daffodil bulbs, and the adults are nectar feeders.

subsoil parts of plants, but the gardener is likely to be oblivious to nearly all of them. We might be better off that way.

Crown Feeders

The soil-level portion of a plant between its roots and stem is the crown. It is the vital link between the nutrition-providing roots and the growing part of the plant. Because of this, crown-feeding insects can be especially devastating to a plant's health. In terms of tissue, there does not really seem to be much difference between the crown, the stem, or the root mass. One suspects that some insects feeding on swollen roots might just as easily migrate upward to tissue found in the crown, and that some insects feeding in the basal stems might just as easily migrate downward. The crown seems to appeal to some insects that apparently do not want to get their tiny feet dirty, so to speak. A case in point is the carrot rust fly. Larvae of this fly live in the swollen roots of carrots and parsnips, but they also live in the thickened crowns of celery. The strawberry crown borer belongs to a group called the clear-winged moths, which have a number of related species that bore into plant stems. There are gall-forming flies and wasps that create galls at the crown level of plants. In truth, these insects are relatively uncommon, but they are related to other gall formers, which number in the thousands, that attack stems, leaves, and flowers of almost any plant imaginable.

Stem, Branch, Shoot, and Trunk Feeders

The difference between stems, stalks, branches, twigs, shoots, and trunks may not always be clear, especially in the early stages of growth. In these stages, there is little difference between herbaceous plant stems and tree stems. Basically, a stem is the part of a plant that bears a plant's leaves and flowers. By this definition, essentially all flowering plants have stems (even if they are hardened, such as a tree trunk). At maturity herbaceous plant stems and woody plant stems change in construction, and it should come as no surprise that a wide variety of insect species take advantage of these altering nuances, some segregating themselves at different heights or depths in tree trunks, others in different diameters of branches or twigs. Because of this degree of preference or ability to specialize, stem resources can be divided up among many diverse families, genera, and species of insects. Thus, a single plant, at various times in its life, might be home to many different kinds of insects.

Several generalizations can be made about stem feeders, and these categories may help define the insects a gardener might find on plant stems. Although there are many different species of insects that feed externally on plant stems, there is relatively little diversity in the orders that do so—almost all belong to the group called plant bugs, Hemiptera (as noted below). These external stem feeders, with few exceptions, are sap-sucking insects during their entire life and have no way to burrow into a stem, no matter how soft it might be. (As an aside, a gardener may find insects on the stems of plants, but in most cases these are simply resting or traveling—their real interest is a search for better feeding grounds.)

In contrast, internal stem feeders all have well-developed mandibles used to chew up and down or around plant stems. If these larvae tunnel longitudinally through stems, they are called stem, shoot, or wood borers. If these larvae tunnel around a stem (or in most cases the adults cut a groove around the stem before laying eggs), they are called twig girdlers. Many diverse species are borers, and the group is composed largely of beetle and moth larvae. Some wood wasp larvae also infest stems.

STEM FEEDERS (External)—Perhaps the most notorious external stem-feeding insects are the cutworms, which are moth larvae. The category of cutworms is comprised of many species, all of which feed at night and attack tender, young seedlings. This is one of the few external stem-feeding groups with mandibles, which they use to cut through seedling stalks at ground level. Cutworms may also attack roots. These insects mostly attack the seedling stage of a plant, which is neither particularly well developed nor hardened off. In this respect, cutworms may be thought of as intermediate between true root feeders and true stem feeders.

Aphids feed on a rose stem. These insects, perhaps the garden's quintessential scourge, are sap feeders that drain fluids from their host.

Other than cutworms, the main external feeders on plant stems are the homopterous members of the order Hemiptera, including soft scales, armored scales, mealybugs, aphids, leafhoppers, spittlebugs, treehoppers, planthoppers, the psyllids (or jumping plant lice), and woolly adelgids. All these insects are sap feeders, and none feed directly on external plant tissue. Many times the feeding appears to have no effect on a plant (although plant diseases may be introduced via sap feeding), but heavily infested plants may show stem dwarfing, stunting, or wilting.

STEM BORERS—Stem borers are insects that bore through the internal structure of plant stems, twigs, and small branches. The eggs of stem borers are either inserted directly into the stem tissue by the adult insect (sometimes using its mandibles to cut a hole) or the eggs are laid externally, hatch, and the larva burrows into the stem. Either way, once the larva is in a stem, it remains there until it matures, at which time it either pupates and emerges as an adult or it burrows out of the stem and pupates in the ground or some protected place.

Stems of herbaceous plants and shrubs are most likely to be bored by larvae of beetles and moths and to a lesser extent by wasps and flies. The bronze birch borer, oak twig girdler, and apple tree borer are fine examples of beetle twig borers. They are also called metallic wood-boring beetles or flat-headed borers, and as we shall see, these insects are also common wood borers. Among moth larvae, a few pyralid moths range in appetite from grass stems and melon vines to woody twigs. Included here are the self-defining walnut girdler, plum borer, and pine moths. The well-known squash vine borer is a member of the clear-winged moth family. A very few wasps, including stem sawflies and wheat jointworms, bore into the stems of grasses. Flies are poorly represented among insect stem borers—they seem to spend most of their energies feeding at root level. Frit flies have larvae called grass stem maggots, and some ephydrid fly larvae bore into stems of aquatic plants as well as cereal grasses.

SHOOT BORERS—This group is just a refinement of the stem-borer habit in which a larva bores into the succulent, apical new growth of a plant. Whereas boring into woody stems and twigs may have relatively little effect on the growth of a plant, boring into a vigorous shoot almost certainly kills it.

The most common tip or shoot borers are moth larvae, which cause damage to terminal shoots of coniferous and deciduous trees. Pine tip moth, pine shoot moth, and peach twig borer are examples of moth larvae named for their mode of living.

WOOD BORERS—Wood borers are simply stem borers that attack tree trunks (that is, woody stems). The most likely insects to be encountered in wood are beetles. Indeed, several families are specialized wood borers, including the flat-headed borers. Larvae of these beetles feed on the wood of dead or dying trees or shrubs. Another common family of beetles contains the long-horned beetles, also called round-headed borers, which infest many different kinds of trees, but prefer those that are freshly cut (or fallen), weakened, or dying. Larvae of these beetles feed variously in dead, dry wood; decaying wood; or under the bark of dead or dying trees.

The bostrichids, commonly known as branch and twig borers, contain species that attack living trees, dead twigs, branches, and seasoned lumber. The bark, timber, or engraver beetles excavate galleries under the bark of trees. In this group, the adult beetle chews its way under the bark, where it deposits its eggs. The larvae fan out and create elaborate patterned tunnels, from which new adults emerge through the bark. The external effect looks as if a shotgun had been discharged. Depending on the species, these beetles attack dead logs, recently cut trees, or living trees of both coniferous and deciduous species.

Although many different beetles are found in wood, a gardener is more likely to find the best-known wood-feeding insect, the termite, or the slightly less well-known carpenter ant. Both of these groups of insects provide a valuable service by converting dead trees to substances that can be used by plants. Of course, when that dead tree is a piece of lumber in your house it is much more difficult to retain a detached perspective.

One last group of wood feeders includes the horntails and wood wasps. Some horntails reach nearly 5 centimeters (2 inches) in length. Their larvae feed in the wood of incense cedar, maple, elm, beech, and many other deciduous trees. The wood wasps similarly bore into dead and decaying deciduous trees.

GALL FORMERS—Stem gall formers take stem feeding one step further: they cause their host plant stems to swell. By doing so, they not only increase the amount of tissue available to feed on, but they also may increase its nutritional quality. The addition of extra layers of plant tissue may also provide some protection from the parasitoids that are always looking for something to parasitize.

The process of gall formation is basically restricted to a few main insect groups, the most diverse of which is the gall midge family. Different species in this family attack an extremely wide variety of plant hosts and collectively attack all potential parts of a plant. A few gall-forming flies may also be found in the fruit fly group, which causes stem and shoot galls on a much more limited array of plants than gall midges, but they are especially attracted to members of the aster family.

Gall wasps, or cynipids, are a second important group of gall formers that cause stem galls on oak, rose, cinquefoil, greenbrier, blackberry, and various asters. A few other wasps such as sawflies cause stem galls on willow, poplar, and honeysuckle, and a pteromalid wasp causes galls on blueberry.

The remaining stem-gall-forming insects are dispersed among a few other orders including buprestid beetles, which cause galls on stems of blackberry and raspberry, and weevils, which cause shoot galls on stonecrops. A few gall aphids attack conifer stems and a few moth larvae cause stem galls on goldenrod.

Leaf Feeders

In terms of available food mass, leaves are probably the most conspicuous and well-used parts of a plant fed on by herbivores. Not

A common form of larval concealment for plant feeders is living inside a gall, which is essentially a plant swelling. The white balls shown here are caused by gall fly larvae that live in the branchlets of a bald cypress. Galls come in many sizes and shapes. They are caused either by chemicals injected into the plant when the adult insect lays her eggs or by the reaction of the plant to larval feeding.

only is there a tremendous diversity of insect groups that feed on leaf tissue, but these insects use leaves in a number of different ways. The two basic methods of leaf feeding are chewing and sap sucking, but these can be classified into categories such as free-living, rolling, mining, and galling.

If a gardener is familiar with any insects at all (other than butterfly adults), it will most likely be the leaf feeders. They tend to draw attention to themselves by their feeding habits and the apparent damage they cause to some of our garden plants. But we gardeners are not alone in our ability to detect (or not detect) these feeders based on our observational skills. Interestingly, the habits, shapes, and colors of many leaf feeders are now known to reflect, at least in part, their palatability to certain predators, especially birds. The theory goes that if an insect is a sloppy feeder that produces big holes, jagged edges, and irregular gouges on leaves, it is probably not tasty and will be avoided by birds. Birds, you see, are visually alerted by foliage that has been damaged—it indicates a potential juicy larva for dinner. Thus, insects that taste bad (which is usually derived from chemicals in the plant upon which they feed) tend to be less concealed and less tidy in their method of leaf eating than their more tasty counterparts. In contrast, tasty insects tend to be much more careful in their feeding habits. They create smooth, regular feeding edges, they often migrate to an undamaged leaf on which to rest, and they sometimes cut off the partially damaged leaf they just fed on so as to leave no visual cue for birds. If you are to be a gardener, it would seem, you must be slightly smarter than a bird to keep up with the insects.

Because there are so many different kinds of leaf feeders, I will mention only a few examples of what might be encountered in the garden. In general, the following insect orders (and their feeding life stages) are almost entirely herbivorous and will commonly be found attacking plant leaves: Hemiptera (true bugs, scales, aphids, leafhoppers, whiteflies, and cicadas; nymphs and adults), Lepidoptera (moths and butterflies; larvae), some Orthoptera (grasshoppers, katydids, and walking sticks; nymphs and adults), and Thysanoptera (thrips; larvae and adults). The large orders Hymenoptera (wasps, bees, and ants), Diptera (flies), and Coleoptera (beetles) have such a wide variety of feeding habits that it is difficult to say anything precise about them, but they all include a number of leaf feeding insects as noted below.

FREE-LIVING—The free-living feeders are those that roam about eating leaves without doing anything special to them. Most butterfly and moth larvae, for example, feed in this way. Many free-living groups of insects are named for their basic feeding preferences: the leafhoppers, for example, or the leaf beetles. Different species in these groups feed on almost any plant imaginable.

LEAF ROLLERS—If you are a leaf feeder, especially one with no special protection, one way to provide a little cover or a hiding space is to encase yourself partially or completely in leaves. Leaf rollers do

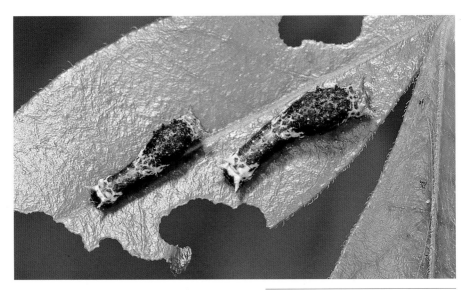

Young swallowtail larvae feed on a citrus leaf—or are these bird droppings? Appearing to be something you are not confuses predators.

Puss moth larva feeding on birch leaves. This tidy feeder is well camouflaged. The larva is probably tasty to birds and so is careful not to reveal itself or its feeding pattern. Because birds can cue in on irregularities in leaf margins, the insect that leaves no evidence of its work is the insect that lives to breed.

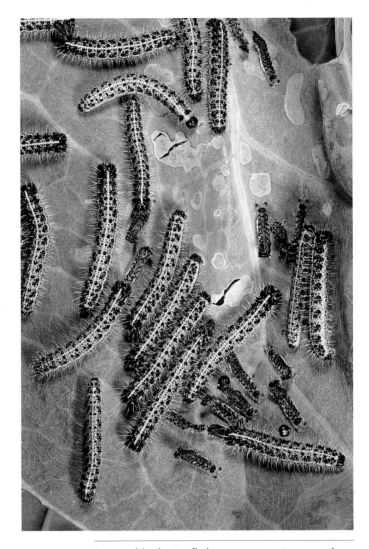

Large white butterfly larvae aggregate on a cabbage leaf. They have no reason to be tidy. The vivid coloring and number of larvae indicate that this species avoids bird predators by tasting bad. If one larva serves as an unpleasant learning experience for a young bird, the remaining larvae will survive as a result of their warning coloration.

this. Essentially, leaf rolls are produced by a larva that pulls leaves together and ties them with strands of silk. Some of these rolls are open at one end so that the larva can move out to feed, then withdraw for protection. People who study leaf-rolling insects classify them into such types as folders, tiers, and webbers. There is a category, however, called ugly nest builders that might not be so self-evident. This category is designated simply because the rolled up leaves, webbing, dead and dying plant parts, and associated insect waste products (frass) tend to create an unkempt mess. An example of this might be the common bagworm, which is most often noted as little brown cones hanging on evergreens. These bags are a compilation of twigs, silk, and dried leaves. The moth larva drags the bag around, emerging from time to time to feed on the tree in which it lives. The fall webworm also might be called an ugly nest builder; it webs up whole branches, leaves and all, which, technically speaking, would make it a branch roller. Sometimes it is better not to dwell on what things are called and simply to accept the way things are.

By far, the most common insect rollers are moth larvae, and the behavior is found in many different families too numerous to mention. A few additional leaf architects may be found among the leaf-rolling weevils, the web-spinning and leaf-rolling sawflies, and the leaf-rolling grasshopper. Except for moth larvae, none of these insects are common in the average garden.

LEAF MINERS—If you feed on leaves exposed to the elements, another way to hide yourself is to sandwich your body within the leaf itself. Obviously, this is a tight fit, so you would have to be a two-dimensional larva. Even so, many insects have adapted to feeding between the outer and inner (epidermal) layers of a leaf, and when they do so they produce a slightly raised, translucent mine visible on the leaf's surface. There are many leaf-mining insects, and again the prize for employing the behavior might best be awarded to various moth larvae.

At least three other orders of insects have adapted to this lifestyle, even though their larvae are usually thought of primarily in other contexts. For example, the flat-headed borer group is one, and the leaf-feeding beetle group is another. Weevils also mine leaves. The wood-boring beetle family is primarily associated with trees and shrubs, whereas the leaf-feeding and weevil groups are associated with grasses, herbaceous vegetation, and sometimes trees.

The larvae of a few fly families mine leaves, but the habit is particularly common in a group called, aptly enough, leaf miner flies.

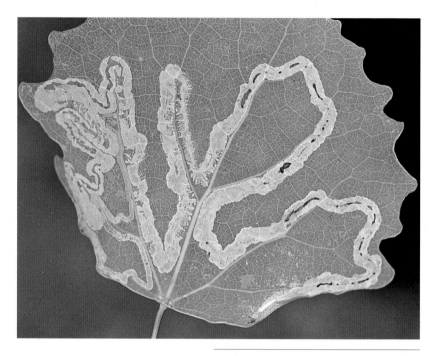

Leaf mines come in different patterns and are caused by the larvae of insects in several different orders. This is an example of a serpentine mine. The larva tunneled just beneath the surface layer of the leaf in a meandering pattern. Other larvae make blotch mines that show up as a solid circular pattern of white on the leaf's surface.

These flies attack a wide variety of plants and are found everywhere. A few families of wasps are leaf miners, but they are not common or widespread.

GALL FORMERS—As with stem gallers, leaf gallers fall primarily into two groups of insects: the gall flies and the gall wasps. The flies attack many different families of plants, whereas the wasps primarily gall leaves of oak and wild rose. Some fruit flies cause leaf galls on members of the aster family. Few other insects are leaf gallers. Among the true bug families, some spittlebugs cause galls, as do many species of jumping plant lice (on buckthorn, rushes, and hackberry) and aphids (on poplar, sumac, and witch hazel). Some species of sawflies cause leaf galls on willows or poplars.

Flower and Bud Feeders

Flower feeders, in general, do not feed exclusively on flowers and flower buds except in special ways (as detailed below). That is, most insects chewing or sucking on flowers are simply leaf or stem feeders that have found a particularly succulent morsel and paused to feed on it. They may first have been attracted by a fragrance, nectar, or pollen. Some insects, however, are entirely dependent on the floral structure, and those are the ones I highlight.

BUD FEEDERS—One of the most famous of all bud feeders is the object of song and legend: the boll weevil. An adult female of this beetle punctures the bud surface of cotton with her mandibles and then inserts an egg into the hole. Attacking the bud causes the flower to abort, or at least reduce its seed production. The boll weevil also attacks okra, which, in my scheme of things, makes it a most valuable insect. A lot of how we think about insects has to do with perspective. I don't think so kindly of the strawberry weevil, which attacks the flower buds of strawberry in the same way as the boll weevil.

NECTAR FEEDERS—Many, many adult insects, even male mosquitoes, feed on nectar. Nectar is produced by glands located in the flowers (nectaries) or on leaves, the bases of leaves, or calyces or bracts (extrafloral nectaries). Flower nectar has only one known function, to attract and reward pollinators, most of whom are insects. It would probably not be an exaggeration to say that almost any insect that can fly, crawl, hop, or flop to a flower would do so if it could. However, the only insects that have developed a total, irrevocable, and absolute dependence on nectar (which they modify into honey) are the honey bees, of which there are several species native to the Old World. Even bumble bees primarily do not use nectar to feed their young: they use pollen. Extrafloral nectaries appear to allow plants a bit of self-defense, by providing a source of food for parasitoids and predators, which may help the plant rid itself of herbivorous insects. Some species of ants are also attracted to these nectaries and guard the nectar, keeping herbivorous insects away from the plant as the ants themselves take full advantage of the plant's resources without harming it in any way.

POLLEN FEEDERS—Having given little credence to nectar as a necessary food source for any one group of insects, the same cannot be said for pollen. We seldom think of insects as pollen feeders, except perhaps as they pass through on their way from nectar source to nectar source. But many insects, including predators such as ladybird beetles, are pollen feeders. Bees, comprising about 20,000 described species in the world, are completely dependent on flower pollen, which they collect to feed to their young. (There are a few parasitic bees, but they are virtually insignificant in the total number.) Bees and plants (that is, their flowers) have a mutualistic relationship in which both organisms benefit from each other. When plants benefit, by being pollinated, so do all life-forms on the planet. Bees are among those hard-working, largely ignored—or even feared—insects that help make our lives more livable.

GALL FORMERS—Gall-forming insects attack flowers at various stages from bud to fruiting. (We will look at seeds and fruits in the next section.) Flower buds are a specialized environment for a number of insects, although again, we rarely think of buds as such because we rarely experience bud damage in our garden—that we know of. Buds are mostly small and inconspicuous, so there is not much to miss. Gall formers also attack the receptacles of composites; even when in full flower, it takes a discerning eye to see a tiny disturbance in the head of an aster. Perhaps the most common bud and flower gall formers are the gall flies, which use many different species of plants as hosts. Some fruit flies are particularly attracted to the receptacles of the aster family.

Other than flies, there are not many common gall formers on garden flowers. A few gall wasps cause galls on flowers and buds of oak and rose. A few weevils and bugs cause flower galls, and an entire group called fig wasps causes galls in the flowers (within the receptacle) of figs. These fig wasps are so tiny, however, they would rarely be seen, even by gardeners in subtropical climes.

Seed, Pod, and Fruit Feeders

In the botanical sense, a fruit is a mature vessel (ovary) that contains the seed (or seeds). Therefore, such diverse appearing things as an avocado, a tomato, a pineapple, and a dandelion seed (with its parachute) are considered fruits. For a student of insects, this concept is rather too broad, because different insects have specialized to feed on various parts of the so-called fruit. Some limit themselves only to the seed (ovule) and feed internally, whereas others feed on the ovary wall (pericarp), which is the edible part of most fruits. Some insects restrict themselves to the receptacle, which is what supports all the seeds (the fruits) of a sunflower. Similarly, in the case of a single strawberry, what we eat are all the fruits and their seeds that reside on a single receptacle, and the receptacle is what an insect would eat. We generally call the whole thing a fruit. As an entomologist, I may be off in my use of floral terminology, but I will do my best to convey those parts of a seed, seedpod, or fruit that are fed on by specific insects. In the case of these structures, it becomes somewhat difficult to classify the type of damage, whether boring, chewing, mining, or galling.

SEED FEEDERS—Because they are a highly nutritious source of food, seeds are fed on by many different sorts of insects. Some specialize in unripe or developing seeds, others prefer the mature, dry seeds. The seed wasps, an ill-defined category composed of species and genera from various families of chalcid wasps, prefer developing seed embryos. (The groups to which these wasps belong are normally thought of as parasitoids of other insects.) Some of these wasps specialize in seeds of leguminous plants; others in conifers; and some in rosaceous plants such as rose, almonds, peaches, and apples. These appear to be the most commonly attacked hosts of seed-feeding wasps. All these wasps are internal, with one larva feeding inside one seed or ovule.

Insects that specialize in mature, or dried, seed are mainly the larvae of beetles and moths. The snout beetles (also known as weevils) are exemplified by the hollyhock weevil, which has a snout (an elongation of the head with the mouthparts at the tip) that is much longer than the entire length of the beetle. Members of this group also feed on the seeds of oak, hickory, and wheat, corn, and rice. The related group of beetles called seed weevils (not true weevils) commonly feed on dried beans.

Several well-known moth larvae attack seeds, one of which animates the so-called Mexican jumping bean. In this case, the larva throws itself against the sides of a bean it has hollowed out and causes the bean to jump. The larva of the yucca moth feeds in yucca seeds that are created when the moth purposefully pollinates yucca flowers with specially modified mouthparts. The yucca has no other known pollinator, and without the moth, seeds would not form. Feeding the moth larvae is small price to pay for such dedicated pollination service.

FRUIT AND POD FEEDERS—A wide diversity of insects are fruit or pod feeders, but the primary groups of such insects are flies, beetles, and caterpillars. Defining their specific types of feeding (for example, chewing, boring, mining) becomes somewhat difficult. If larvae feed just under the skin of a fruit (which includes what most of us think of as vegetables), then they are called fruit or pod miners. The apple maggot is one such insect. If larvae feed in the flesh of a typical fruit, such as an apple, then it is considered a borer, even if it only hollows out a depression on the fruit's surface. A list of fruit-boring insects reads like a who's who of garden plunderers: cherry maggot, fruit moth, fruit flies (of various persuasions), plum weevil, pickle-worm (it should be called the cucumber worm), lesser apple worm, grape berry moth, tomato fruit worm, apple curculio, quince cur-culio, melon worm, and so forth.

If a larva feeds in the seedpod of a plant, on tissue of both the inner pod wall and sometimes the seeds themselves, then it might be called a pod feeder: it is living in a room composed of plant tissue and simply chews its way around the room eating whatever is tasty. A number of moth larvae behave in this manner.

GALL FORMERS—It is sometimes difficult to pinpoint exactly what part of a fruit a gall-forming insect feeds on because the entire fruit, or at least parts of its interior, become distorted during the galling and feeding process. Because of this distortion, it becomes a moot point as to whether the distorted object is an early or late flower bud, or perhaps an immature seedpod. Among insects, one family stands out as the gall former to the greatest variety of fruits: the gall midges. These flies create galls in the seeds of such common plants as verbena, butterfly bush, and sunflower; in the fruit of pear, cactus, blueberry, and violet; and in the seeds of meadow rue, cleome, and hibiscus. The remaining insects that cause seed and fruit galls are fairly uncommon. At least one weevil causes fruit galls on veronica. Some gall wasps cause galls both in the seeds and cups of acorns, and some related seed-feeding wasps cause galls in seedpods of legu-minous plants.

Special Types of "Plant Feeders"

To say that insects feed on plants is a simple-sounding statement. To say that they feed on plants in many different ways, as I have tried to demonstrate, complicates the picture only slightly. To complicate the picture hopelessly, one need only delve slightly further into special types of insect-plant associations, of which there are many. Next, I point out a few ways in which plant feeding becomes not only more complicated, but also less obvious when examined more closely.

Inquilines

When herbivorous insects feed, they sometimes inadvertently create spaces for other insects to live and feed as well. When a gall midge, for instance, induces the formation of a gall, much of the plant tissue is unused by the fly larva itself. This excess tissue can be used by insect freeloaders—sometimes called guests—which rarely cause the galling insect harm. These guests are technically referred to as inquilines. Gall-forming insects create the most space for such guests, but any of the leaf-mining and leaf-rolling insects create food sources as well. I have read of a fly larva that lives in the froth of spittlebugs, where it obtains both shelter and food. Inquilines, it seems, are not nearly as demanding as many of the house-guests I've known.

Fungivores

Insects that specialize in eating fungi are called fungivores. The majority of these creatures seem to be flies and beetles. Simply listing their common names pretty much tells the story. There are fungus flies, fungus beetles, minute tree fungus beetles, pleasing fungus beetles, handsome fungus beetles, silken fungus beetles, round fungus beetles, shining fungus beetles, tooth-necked fungus beetles, hairy fungus beetles, and fungus weevils. These are all passive fungi feeders that live off materials they find in nature. Some insects, however, cultivate their own fungus, the most notable of which are fungus ants. These ants cut leaves, chew them up, and inoculate them with fungal spores taken from their own fungus gardens. In addition to some ants, a few termites and a beetle or two are the farmers of the insect world.

Plant-feeding insects use different devices to hide from their enemies, the predators and parasitoids. In addition to hiding in galls, leaf rolls, or leaf mines, some insects create more imaginative cloaking devices. These vulnerable, soft-bodied, immature spittlebugs cover their bodies with a mucilaginous, frothy substance extruded through the anus. Living in bubble-wrap provides both concealment and a protective cover of sticky, spitlike goo.

Detritivores, or Saprophages

Some insects are feeders in dead or decaying plant tissue, dead animal materials, or dung. Technically speaking, these leftovers might not be considered plants, but the distinction becomes somewhat arbitrary when one enters the world of the general feeder of all things dead and decayed. The dung of plant-feeding mammals, for example, contains a large amount of plant fiber and cellulose—enough leftovers, as a matter of fact, for a never-ending meal for insects that like that sort of thing. When decaying plant parts mix with decaying insect or bird parts, who knows what is being eaten? The filtering of all this dead and dying material through the guts of endless numbers of insects and bacteria releases nutrients back into the soil for the growth of yet more plants. A large number of different insects eat decaying material, prefer it even. Who are we—who eat snails, fungi rooted out by pigs, and the well-rotted excretions from a cow's udder (cheese)—to comment unfavorably of the fact? We should be happy, after all, that something does the dirty work. Among the insects to give thanks for (in this department) are many sorts of beetles, termites, flies, springtails, and cockroaches. Yes, even cockroaches have a purpose in this world.

Omnivores

Many hundreds of thousands of insects are fairly easy to classify with regard to feeding behavior. Even the detritivores are relatively predictable when it comes to what they eat. There is a type of feeding behavior, however, that covers all bases, an omnivore: an organism that eats almost anything. A number of insects might be said to be omnivorous. Many ants, for example, scurry about the environment scraping up bits of this and that, sometimes killing other insects, sometimes taking plant parts, sometimes dismembering dead insects. The mole cricket feeds extensively on grass roots, but also eats earthworms and other insects. Many insects commonly thought of as predators are somewhat omnivorous in that they will eat pollen and nectar, especially when prey items are rare. Some predators also obtain water from plant sap. Even some unsuspected insects are known, on occasion, to turn the tables.

Once, while viewing a living exhibit at the insect zoo at the National Museum of Natural History, I saw a cage containing a poor, defenseless katydid, a plant feeder, that had been offered up as the daily meal for the insect predator of all time, a praying mantid. However, the mantid was hanging lifeless in one foreleg of the katydid, which was dining on the head of its supposed killer. As we shall see in the next chapter, with insects not everything is as it might seem.

As you may have gathered, even single insect groups, such as beetles, have endless ways of feeding on plants. Different families within an order can feed on roots, stems, leaves, flowers, fruits, or dead tissue. Sometimes different stages of the same species can feed on roots (as larvae) or leaves (as adults). Within the same order there can be free-feeders, gall formers, miners, rollers, or borers. It is difficult, therefore, to make comprehensive statements about any one group of insects that hold up under close scrutiny. It seems that one can always find an exception to a statement. In fact, from time to time entomologists have heated debates about insect feeding. This is especially true if the behavior appears to be contrary to what is currently believed. For example, when plant-feeding wasps were found in a family of wasps known only for its parasitic habits, it took several years and much evidence to demonstrate that, indeed, plant feeding has evolved in groups of insects that are essentially all parasitoids.

I point this out not to confuse the issue, but to suggest that it is often difficult to categorize the kinds of insects you might find in the garden. Even if the insect is captured on a plant, there is no easy way to state what that insect is—plant feeder or predator—without some observation and a little patience. In the next chapter, we will explore how the plant-feeding insects interact with the predators and parasitoids that attack them.

7

The Interactions of Insects with Each Other

In the last chapter, we learned that insects and plants interact in a great many ways. Thus, it should come as little surprise that insects interact with each other in an equally vast number of ways. Interactions between insects, however, may become extremely intense and are more complicated than the relationships between plant-feeding insects and their hosts. The relationships are more intense because both participants are active, and as anyone who has seen nature shows on television can tell you, watching predators pounce on their victims is much more intense than watching a cow mow down a field of grass. Not necessarily more pleasant, mind you, but certainly more visceral. Then, too, the relationships can become more complex because insect-eating insects may have several layers of specific host relationships to wade through before they find the precise host they are looking for.

Although not apparent to most gardeners, our backyards can be as vicious as any imagined jungle. There is a whole lot of pouncing going on as one insect species attempts to promote its kind at the expense of another. These exploits can be as openly gruesome as the praying mantid bagging its grasshopper supper, which it eats alive, or as apparently tranquil—although infinitely more insidious—as a tomato hornworm walking blithely about as dozens of tiny parasitic larvae consume it from the inside. The praying mantid is a miniature example of a predator, more commonly known to us in the form of lions, tigers, wolves, or foxes. The hornworm invader, technically called a parasitoid (or sometimes not quite correctly, a parasite), is less well known to us because we see no easily recognizable counterpart among any of the large animals we know.

Animals that eat insects are entomophages and their behavior is entomophagy. The more general terms *zoophages* or *carnivores* are also sometimes used. (Entomophagy is a specialized category of the carnivory we examined in chapter 4.) When speaking of entomophagy, we are basically talking about two types of insects: predators and parasitoids. In this chapter, we look at the basic ways in which insects eat other insects and the groups of predators and parasitoids likely to be found in the garden.

Predators

Many of us are familiar with predators based on personal observations of our killer cats and murderous dogs. Although these pets may be well fed, they still have a natural instinct that allows them to pounce on the unsuspecting bird or rabbit when the occasion permits. It comes as no surprise that their relatives in the feline and canine worlds are excellent predators as well. We tend to think of predators as those animals that stalk or lay in wait for their prey, pounce on it, and eat as much of their victim as they can possible choke down. It is not a pleasant concept, of course, but except for the stalking, it scarcely differs from our own behavior when it comes to the Thanksgiving turkey.

A great many insects are excellent hunters and predators. If they were more conspicuous—say, the size of a lion—the human race would most likely be nonexistent. The adaptations of insects to the predatory life are so diverse and marvelous, in a sadistic manner of speaking, that it seems unlikely anything could escape their ravenous appetites. A single example might demonstrate such a statement. As an adult, the swiftly darting dragonfly is among the world's greatest flyers—it puts any bird to shame as it cruises about at breakneck speeds, snatching other insects that are in flight or at rest. It then chews these insects with its powerful jaws. As an immature insect, however, the dragonfly lives underwater. The nymph has a completely different set of mouthparts, with a set of pinchers at the end of a hinged plate that normally rests folded under the head. When a prey item approaches, the nymph shoots out the plate, grabs the prey with its pinchers, and pulls it back to the mouthparts at the base of the head. Here the prey is ground into paste by the jaws.

You generally may be familiar with such insect predators as the praying mantid, ladybug, antlion, and perhaps the green lacewing, however, the insect world teems with predatory forms. Although this is a book about insects, it should also be mentioned that spiders are considered among the most important predators likely to be found in the garden. As spiders are studied more intensely, we are beginning to learn that these often-maligned creatures are far more common and significant than has previously been known.

We will examine some of the main insect predators found in the garden, proceeding from the most conspicuous (and usually largest) to the smallest and least likely to be seen.

Praying Mantids

The largest insect predator a gardener is likely to find is the praying mantis, or mantid. There are fewer than a dozen native species in the United States, and the two most common mantids to be seen in the garden are introduced from other parts of the world. One is the Chinese mantis, the other the European mantis. Praying mantids are sedentary hunters that wait in ambush for their prey to come along. They attack victims at lightening-fast speed, using their raptorial front legs to grasp the hapless prey. Mantids will attack essentially any living insect and have been known to eat frogs and lizards. In spite of their pugnacious habits, however, they are one of the relatively few insects that children can tame and keep as pets. Their antics are comical and are accentuated by their large eyes and an ability to rotate their heads to follow the movement of their captor. The stories of females mantids eating their mates after copulation

Predator eats predator. Here, the famed
praying mantid feeds on its prey, a hunting wasp. Highly touted as a beneficial
garden insect, the mantid would eat children if it could manage. Although the
mantid is a wonderful insect for the garden, its prowess is indiscriminate—it eats
butterflies, too.

appear to be exaggerated, and the behavior may be common only when these insects are kept in captivity.

Katydids and Jerusalem Crickets

The second largest insect predators the gardener will find (although uncommonly) are katydids and the Jerusalem cricket. Both insects are primarily herbivorous, but when presented with an opportunity, they will eat other insects. In fact, an alternate name for the Jerusa-

A katydid sits on a geranium flower. These relatives of grasshoppers are opportunists, feeding on both plants and insects. They simply eat any slow-moving insects they find associated with the leaf they are eating.

lem cricket is potato bug, in reference to its fondness for the vegetable. In truth, these creatures might be thought of as omnivorous or at best opportunistic predators, basically eating anything that strikes their fancy.

According to my entomological colleague David Nickle, katydids will prey on slow-moving insects such as caterpillars and on sessile stages of insects such as eggs and molting stages. Katydids are active at night and are not commonly noticed in the treetop perches where they normally live. It is unlikely that a gardener would even know they were around except for their nighttime rasping or if they were seen flying into the porch lights in the evening. Members of the closely related cricket group behave much as do katydids, according to Nickle. This includes both the ground-dwelling crickets (house and field crickets) and the tree crickets. Both feed on other insects when the occasion arises.

Jerusalem crickets are well known to gardeners in the western United States as one of the ugliest creatures they are likely ever to encounter in real life. With relatively huge (up to 5 centimeters [2 inches]), bulky bodies, these wingless, ponderous creatures are found under rocks and boards during the day. Like their cousin, the more glamorous katydid, Jerusalem crickets roam the nights in search of food and attack slow-moving and more helpless members of the insect clan.

Dragonflies and Damselflies
Dragonflies and damselflies are predators in both immature and adult stages. As adults, these creatures are voracious feeders on mosquitoes and other small flying insects, and as nymphs they feed on anything that lives underwater, including small fish and tadpoles. Chances are, if you have a garden pond, these insects will find it and establish residence.

Wasps and Ants
Some ants and almost all solitary and social wasps are predators, but in several slightly different ways than we normally think of

predators. In this case, the adults stalk and catch the insect prey, but they seldom eat the prey themselves. Instead, they offer it to their larvae, which are concealed in some out-of-the-way corner of the garden.

Several of the more commonly seen groups of predators in the garden are the hornets, yellow jackets, and paper wasps. These are all social wasps that live in colonies varying from a dozen (paper wasps) to thousands (hornets, yellow jackets) of individuals. The social wasps constantly scour bushes, trees, shrubs, perennials, and

An adult male dragonfly rests. These predators maintain territorial posts, where they keep watch for prey such as mosquitoes, male intruders who violate their space, and females who might be in the mood for companionship.

182

Yellow jacket adults feed on sugary substances (in this case an apple). They also
eat proteins found in other insects (and
stray hamburgers) and feed this masti-
cated protein to their young. Paper wasps,
hornets, and yellow jackets are especially
important in reducing leaf-feeding caterpillar populations.

other vegetation looking for all kinds of insects—they are especially fond of caterpillars—which they chew up and feed to their young. This is a form of progressive feeding, where the young are fed as they need nourishment.

Many families of solitary wasps, as the name implies, do not live in colonies. Each female wasp collects specific insects, places one or

A female hunting wasp has captured her prey, in this case the caterpillar of a small moth, and holds it beneath her body in preparation for the descent into her burrow in the soil (the opening to the right). Several larvae will be captured and subsequently placed in an underground cell. Then the wasp will lay an egg on each of them, and her offspring will benefit from her dutiful work.

more in a cell, lays an egg, and then flies away. The larvae are left to eat whatever the female wasp has given them and no more. Solitary wasps are less well known than the social ones, but there are many more thousands of species than the social wasps. Solitary wasps vary in size from those small enough to collect thrips and aphids to those that attack cicadas and grasshoppers. Each wasp species generally specializes in hunting one type of prey, from leafhoppers to others caterpillars, spiders, flies, true bugs, or even other wasps.

Ant habits vary widely depending on the species. Fortunately, we northern gardeners need not fear the tropical army ants that scour their paths of all life-forms, but these ants do have harmless relatives that live in the southern United States and prey on other ants. In our gardens, many ants act primarily as scavengers, feeding on both plant and animal matter. When an ant attacks another insect, it is often a victim that is incapacitated or moribund for some reason other than an attack directly from the ant itself. There are some ants, however, that are entirely predatory on other insects. Most ants live in underground colonies, but some use rotting wood or build carton nests from wood fiber. As with social wasps, ants collect food, which they chew up and regurgitate for their young. Chances are that most gardeners (including myself) will rarely be able to tell one ant from another, and because we rarely see the extensive colonies of ants under our feet, we mostly ignore ants unless they invade the house.

Among the social wasps, solitary wasps, and ants, many different sorts of insect prey species are harvested from the garden. For this reason (and because of the parasitoid forms discussed below), the insect order Hymenoptera is considered one of the preeminent forces for insect population control. Because there are scarcely any negative effects of Hymenoptera in the garden, their presence should be strongly encouraged. In spite of the reputation for stinging (mostly due to hornets, yellow jackets, honey bees, and, in some areas, fire ants), this order is one that the gardener needs as an element of balance in the garden.

Beetles

Among insect predators, more than half are probably beetles of one type or another. This is not surprising, considering that the number of beetle species is believed to be the greatest for any group of animals on earth. When the question of predatory beetles comes to mind, our thoughts most likely drift to the ladybird beetles (or ladybugs). These brightly colored insects are considered a gardener's best friend—they eat aphids, scales, and mealybugs their entire life long, first as ambulatory larval dragons then as puttering adults. There are hundreds of ladybug species in the United States, most of which are predatory, but it should be noted that the Mexican bean beetle and the squash lady beetle are plant-feeding members of the same group. So one should keep an eye out for the few bad ladies in the bunch!

A lesser-known group of predators is the ground beetles. As the name implies, these beetles live at and below the soil surface, and both larvae and adults feed on insects, mites, snails, and earthworms. The adults also climb trees and shrubs looking for prey. These beetles are secretive and are rarely seen, although some are large (more than 2.5 centimeters [1.0 inch]). One species was introduced into the United States to help control gypsy moths. A species of European ground beetle was shown to eat nematodes, thrips, caterpillars, weevils, aphids, fruit flies, silverfish, and slugs (but curiously not slug eggs). Most ground beetles appear to be generalist predators, and not too fussy at that.

A commonly noticed beetle, at least in eastern climes, is the firefly (or lightning bug). Not only do these beetles bring great joy to the evening garden in the early days of summer, they are predators in the bargain. The larvae live on low vegetation and eat smaller insects and snails. Both males and females of the light-emitting species of fireflies (not all produce light) emit flashing lights. These lights are caused by a chemical reaction that takes place in a special organ in the abdomen (oxygen is mixed with luciferin and the enzyme luciferase to release energy in the form of light). Fireflies use these flashes to signal each other from a distance and then to

186

entice each other into the delicate matter of courtship. Perhaps not so delicately, some female fireflies have evolved the ability to mimic the flashing patterns of other species of fireflies. After luring a male of one of these other species to her side, in the promise of an evening's delight, she promptly eats him.

There are relatively few other groups of predatory beetles likely to be encountered by the gardener. Some of the rove beetles and carrion beetles, in addition to scavenging for dead things, also attack a variety of insects, mites, and snails. If you have a pond, you might find predaceous diving beetles beneath the water or whirligig beetles on the water's surface. However, in all likelihood most of these beetles would go unnoticed.

Flies

The average gardener is well acquainted with flies, mostly in a negative context associated with biting, dead bodies, garbage, and excrement—not exactly an advertisement for respectability. It would seem difficult to love a fly, but there is one group for which gardeners have great affinity. These are the flower flies, or syrphids, in which the larvae are predatory. These flies, which are about the size of a small bee, actually resemble bees or wasps, except that they tend to hover in the air more like hummingbirds than do their lookalikes. Adults of these flies are nectar feeders (thus the name flower fly) and are frequently seen alighting on small flowers. The larvae are rapacious aphid feeders.

Robber or assassin flies are less well known than flower flies. They are most likely to be seen by folks who garden in dry habitats, especially deserts of the western United States. These flies are commonly seen at apparent rest in sunny spots on rocks and sticks. These are perches, however, and any insect, no matter how large or fierce, that enters into the fly's territory is likely to be attacked and killed. These flies have much the same effect as praying mantids: they are interesting to look at, but they probably don't do much in the way of decimating the populations of other insects.

Lacewings (Aphidlions) and Antlions

Gardeners are generally familiar with green lacewings (or chryso-pids), which rank right up there with ladybird beetles in popularity. As with the more familiar ladybugs, both the larval and adult stages of lacewings are predatory and eat a variety of notorious insects such as aphids, mealybugs, scales, whiteflies, caterpillars, leafhoppers, psyllids, thrips, mites, and almost anything else they can sink their jaws into. There are actually several species of green lacewings as well as another, less common group, the brown lacewings, which do about the same as the green ones. In the western United States, a small, curiously shaped, ferocious-looking predator, the snakefly, is occasionally seen in the garden. These also feed on aphids and small

An adult flower fly, or syrphid, hovers over a purple everlasting. In flight these flies appear to be small wasps or bees. Flower flies are nectar feeders as adults.

insects. Gardeners who have exceptionally sandy soil may occasionally see conelike depressions in the sand. These are made by larvae of the antlion, often called a doodlebug, which feed on ants and other small ground-dwelling insects that stumble into their slippery-sided pits.

There are hundreds of other predatory species of lacewings and antlions, and the group should be highly prized in the garden. Most gardeners, if they have seen any predators at all, would likely only know about the single green lacewing. Fortunately, we can put all our eggs in one basket—and the rest of the order will still go about its business without a worry.

The larvae of flower flies are voracious predators of aphids. Many insects display different feeding habits between their immature and adult stages.

True Bugs

As we saw in chapter 2, true bugs are placed in a single order of insects. This is among the more confusing groups of insects with which gardeners have to interact. Most gardeners are likely to be more familiar with the homopterous subgroup of the order (aphids, scales, mealybugs), which are all plant feeders, than with the heteropterous subgroup, which has both plant-feeding and predatory members. Gardeners may know of plant-feeding bugs with names such as stinkbugs, chinch bugs, lace bugs, and plant bugs, but this same group also has such sinister-sounding family names as the ambush bugs, pirate bugs, and assassin bugs. All the true bugs, regardless of host, feed by needle-like, sucking mouthparts with which they drain the juices from their host, whether sap or blood.

In truth, several families of bugs that we think we don't want in our gardens (that is, the plant feeders) also contain species that are predators of many other insects. For example, the stinkbug group includes the harlequin cabbage bug as well as the spined soldier bug, which attacks beetle larvae such as the Colorado potato beetle and the Mexican bean beetle. The chinch bug, considered to be the most injurious plant-feeding bug in the order, also contains the big-eyed bugs that feed on mites, aphids, and leafhoppers. A group called the leaf and plant bugs is known for the tarnished plant bug, which seriously damages fruit, but also contains a group of predators that attack plant-feeding stinkbugs, aphids, and syrphid larvae. Before the gardener begins attacking members of these groups, then, it becomes rather critical to distinguish between the good, the bad, and the merely ugly. And because this task is relatively difficult, even for entomologists, it is better for gardeners to simply let nature sort out its own cast of characters, rather than trying to do it themselves.

There are several entirely predatory groups of bugs as well as the heterogeneous families just mentioned. The ambush bugs, assassin bugs, and pirate bugs—as you might guess from their fanciful names—all are killers at heart. They are not too particular about what they attack, either, so they represent a fairly neutral picture in the garden: they kill as many bees, for example, as plant-feeders.

There are a number of other predators in the heteropterous subgroup that can be found living in a functional pond (one that is full of algae and plants, not chemicals). With the exception of water striders, those bugs that skate effortlessly across the top of ponds, the average gardener will probably never see such things as the back swimmers, water boatmen, or giant water bugs (also called toe biters). These voracious predators feed on anything living in the pond and small enough to stab their beaks into, including frogs and fish. Water striders feed primarily on small insects that fall onto the pond's surface.

Earwigs

Earwigs are generally considered to be plant pests, but this is an undeservedly harsh and largely erroneous rap. Although earwigs do eat plant material, this habit appears to be primarily the result of being a fastidious gardener. That is, where garden soil is barren (what neat gardeners call well-groomed), such as in a vegetable or seedling bed, earwigs have little to feed on and so take what is available. Under somewhat more natural conditions (for example, soil covered with humus or mulch), earwigs will more readily find the things they prefer to feed on such as mites, nematodes, insects, and decaying matter. Earwigs are known to actively feed on soft-bodied insects such as aphids and mealybugs. According to the authors of *Common-Sense Pest Control* (Olkowski et al. 1994), earwigs are often blamed for damage caused by snails, slugs, and cutworms simply because they are found near the scene of the crime, where they may just be taking cover for the day. If gardeners are inclined to destroy all apparently harmful insects, it may be that they are doing exactly the wrong thing—destroying the predators that would perform the service at no charge.

Thrips

Because of their small size, 2 millimeters (⅛ inch) or less, thrips are rarely seen by the gardener. Most thrips are plant feeders, but some species are both predatory and fungal spore feeders. Because they are so small, their prey is mostly limited to other thrips and mites. Scarcely any gardener or right-thinking person, for that matter, can tell the difference between herbivorous and predatory thrips. A heavy infestation, which is first indicated by noticeable plant damage, will certainly indicate plant feeders. Believing that predatory thrips exist is one of those leaps of faith that the gardener must make.

Moths and Butterflies

It is highly unlikely that gardeners will find predatory caterpillars in their gardens, but I mention them simply to illustrate that even the most seemingly benign group of insects has its darker side. In the continental United States, the harvester, a delicate butterfly, stalks aphids during its caterpillar stage. Other blues, as lycaenids are called, prey on scale insects. There is also a group of moth larvae that feeds on paper wasp larvae. In Hawaii, there are moth caterpillars that actively hunt small wasps, crickets, and flies. Seldom do we think of moths and butterflies as predators, but with insects almost any sort of behavior is likely to turn up.

Parasitoids

If predatory insects appear numerous in kind and relentless in attack, then parasitoids should be considered more numerous and infinitely more ingenious. Although predators attack their prey fairly viciously and directly, as many a nature show has demonstrated, parasitoids act in a staggering variety of ways, almost all involving stealth and cunning. Many books have been written about parasitoids, but we need not go into vast detail to gain a general idea of how they survive. The difficult part about discussing the group, however, is that their behaviors can become so complicated as to make general, true statements about them almost impossible. I will try my best to be forthright and informative, but to do so I will leave out many of the finer details that would otherwise cause consternation on both our parts.

Although the number of species of insect parasitoids is extremely high, the particular orders of insects that contain them are much fewer than those of the predators. From this viewpoint, then, and because the behavior of parasitoids is relatively more complex than that of the simple-minded predators, I first discuss the types of basic behaviors found in parasitoids and then briefly discuss the principle groups of insects involved.

Primary Parasitoids

The term *primary* is used to define insects that directly parasitize other insects (or spiders, ticks, and mites). Usually these hosts are plant feeders. A typical example of this would be a wasp that lays its eggs inside a moth caterpillar. The wasp simply finds a larva, inserts her ovipositor (that is, the egg-laying tube) into it, and deposits one or more eggs (the specific number depending on the species of wasp). Although this appears straightforward, there are many variations in which different species of primary parasitoids might attack the same species of host. Some wasps attack the early larval stages, whereas others prefer later stages. (Recall that there may be four or five different stages, or instars, in a caterpillar's life.) Some parasitoids lay eggs internally and some lay them externally. Some wasp species complete their entire development within the egg or pupal stage of a host. A few parasitoid wasps lay their egg in the host egg, but it does not develop until the host becomes a larva or pupa. Rarely, adult insects are even attacked.

No matter what the host might be or how or when the egg is laid, the first parasitoid that attacks the host is always referred to as the primary parasitoid.

Secondary Parasitoids, or Hyperparasitoids

The next level of parasitism is secondary, or hyperparasitism (technically speaking, hyperparasitoidism). Hyperparasitoids are so named because they attack primary parasitoids. The phenomenon of secondary parasitism might sound utterly self-defeating—parasitoids attacking parasitoids—but it occurs in one of the most abundant groups of insects and provides opportunities to diversify the environment so that more species can join in the mutual slaughter. Consider a primary parasitoid, for example, a wasp that attacks aphids. A secondary parasitoid would find the parasitized aphid—and only a parasitized aphid—and then oviposit into that aphid by placing its egg on, in, or near the wasp larva that is already inside. Just as you can have hundreds of species of aphids on hundreds of species of

Parasitic wasps (parasitoids) account for some 60,000 known species of insects. Not surprisingly they come in a variety of shapes and sizes. The ichneumonid wasp shown is among the larger species (15 centimeters [6 inches]) and has a remarkably long ovipositor (egg-laying device). Although these parasitoids can penetrate wood with their ovipositor in search of hosts, in this case a wood wasp larva, the ovipositor is too flexible to hurt humans.

plants, you can have hundreds of species of aphid parasitoids and dozens of secondary parasitoids.

Obviously, if you wish to control aphids, having secondary parasitoids is not a good thing because they destroy the parasites that kill aphids. Ecologically speaking, however, there are always fewer secondary parasitoids than primary, so that the gardener need not have an irrational fear of running out of primaries. Besides, things could get worse—or better, depending on how much complication you can handle.

Tertiary Parasitoids

As the name implies, these are parasitoids that attack secondary parasitoids that attack primary parasitoids that attack a host insect. This happens rarely, but it does happen. I could tell you that things become even more complicated, but for the purpose of general knowledge, I think we have about gone where no gardener really needs to go. Be aware, however, that in the realm of biology, even we biologists do not necessarily know what is best. Thus, attempting to intelligently manipulate these systems in the garden is basically not possible, nor perhaps even in our best interest. It is in our best interest to let the bugs work out their own methods of population control with as little interference from us as possible.

Parasitoid Groups

There are perhaps 100,000 to 150,000 described species of insect parasitoids in the world, and estimates of undiscovered species range upward to 1 million. Parasitoidism is an exceedingly common lifestyle, but interestingly only two major orders of insects have evolved to take advantage of it: Hymenoptera (wasps, bees, and ants) and Diptera (flies). (A few other orders of insects have isolated species that are parasitic, but they are essentially unimportant to the gardener.) Of these two primary orders, the most numerous and important, by far, is the Hymenoptera.

WASPS—We gardeners are most familiar with the order Hymenoptera based on our interactions with the social bees (honey bees and bumble bees), ants, and wasps (hornets, yellow jackets, and paper wasps). Our knowledge of the parasitic forms is essentially nonexistent because they are generally tiny, relatively cryptic, and poorly known even to professional entomologists. Wasps that average the size of a grain of rice (or much smaller) scarcely draw your attention and can sometimes appear to be so uncommon as to serve no great purpose. But this viewpoint is contradictory to the ecological facts. These wasps are doing their job, which is finding their host insects and keeping the host population at low levels. It is when this job is

A female trichogrammatid wasp, a minute parasitoid that is less than 1 millimeter (0.04 inch) in length, oviposits into a moth egg that is bigger than she is. The wasps are so small that two or three adults might emerge from each egg.

interrupted (for example, when these few parasitoids are poisoned out), that the host population can grow unperturbed by meddlesome enemies.

Parasitic wasps attack so many different sorts of insects and in so many different life stages that it almost becomes easier to list which groups of insects are not attacked than which groups are. In the garden, the job becomes easier. Most wasp parasitoids attack immature insects; adults are rarely parasitized. (One rare exception to this rule might be the relatively common garden insect, the ladybird beetle.) Therefore, any adult insect seen in a garden will likely not be attacked by parasitoids (they may, however, be parasitized by mites or other organisms). However, any immature life stage is relatively susceptible to attack by parasitoids, including eggs, larvae, nymphs, or pupae. Any insect species commonly found in the garden is likely to have an associated parasitoid or two. Some of the gall-forming insects, for example, may have as many as ten or twenty different parasitoids associated with them, and some of the houseguests (or inquilines) that use their gall.

With such a biologically diverse and enormously large group as the Hymenoptera, it is difficult to provide a more precise categorization than I've given. In the interest of communicating an impression of these parasitoids, without going into endless detail, I perhaps risk implying that the group is not as biologically significant as it is. However, hymenopterous parasitoids are at the forefront of regulating insect populations. If gardeners realize even this bit of hyperbole (after all, I make my living by studying parasitic wasps), then they will realize the impact of enticing a resident population of such creatures into their gardens.

FLIES—If gardeners are relatively unfamiliar with the vast hoards of parasitic wasps, they are clearly ignorant of the far lesser hoards of parasitic flies. Flies attack such insect hosts as bees, ants, psyllids, beetles, moths, grasshoppers, crickets, termites, plant bugs (for example, leafhoppers, treehoppers), earwigs, and rarely other flies. They also attack some rather odd groups of animals including land

and aquatic snails, slugs, earthworms, centipedes, and spiders. A few species of flies, it might be noted with alarm, have the odd habit of hatching eggs in their ovaries; these maggots, empowered with some absence of filial devotion, proceed to eat their mother alive. (Given the choice between such primitive behavior and that of the more elegant hymenopterans, is it any wonder I choose to work with the latter sorts of parasitoids?)

In this chapter I presented some of the ways in which insects interact with each other. Because this chapter is shorter than the previ-

This fly does not appear much different than a house fly, yet its larvae behave entirely differently. This is a parasitic tachinid fly, and the group as a whole is especially fond of the larvae of moths and butterflies. Tachinid flies also attack sawflies, wasps, beetles, members of the order Orthoptera, and the nymphs of true bugs.

ous one, you may have the impression that predators and parasitoids are biologically less complicated and have fewer types than plant feeders. As to the first notion, I would say that insect-insect interactions are somewhat easier to categorize, but only because their natural histories and behaviors vary so tremendously that it is impossible to label them except in the broad sense of "predator" or "parasitoid." In terms of species, these two groups may be fewer in absolute numbers than the plant feeders, but in terms of lifestyle diversity, they are likely equal.

The interactions of insects with each other and with plants represent the primal and eternal elements of a garden. How the gardener views these interactions is as much a matter of awareness as knowledge. In the next chapter, we will broaden these perspectives based on our growing understanding of our gardens' dutiful inhabitants, the insects.

8

Conceptions: A View to the Garden

As we saw in the introduction to this part, when it comes to the question of bee stings, sometimes technical correctness can obscure the heart of a matter. We become bogged down in the attempt to find a clear, concise, correct answer. Perhaps it would be better simply to ignore correctness and be wrong for the sake of communication. But sometimes, in our haste to find simple answers, we become simple-minded and can no longer view life in its true perspective (or "alternative perspective" might be a better way to put it). In this chapter, I attempt to explain the differences between what we gardeners think we need in our gardens (that is, our perceptions of *the way things are*) and what actually takes place in our gardens (the naturalistic condition, or *the way things ought to be*). These are not particularly complicated notions, but occasionally it helps if we step back from our perspective and assess the situation in an open and objective manner.

Human Perception

We gardeners suffer from a bad case of false perception when it comes to designing and building our gardens. We think that because we want something to be so, it will be. We attempt to grow alpine plants where bananas would sprout. We put shade plants in the sun and sun plants in the shade and wonder why neither grows. If aphids appear, we kill them with pesticides and then wonder why there are no butterflies. I know these things because I am a gardener who has tried them all. I ought to know better, but apparently I don't. We gardeners are experimenters, if nothing else, always mindful that something might work if we try it often enough or try it long enough. And sometimes it will. The biological world allows us to succeed in fooling Mother Nature once or even twice, but she will always be herself in the end.

Let's examine the way we gardeners perceive the garden and what we do in its name, while alternatively ignoring the way a garden behaves in the natural world. Then we will examine the natural world and how the garden might fit into it. By comparison, perhaps we might learn a thing or two about success and failure as it relates to our view of insects.

Neutrality, or Ignorant Bliss

When it comes to insects, most gardeners—and I include myself here—will likely have no inkling as to what is generally happening during any given time or season in their gardens. Regardless of what you may think of me, I do not take daily collecting trips into the garden to see what is there. Gardeners generally perceive their garden as a collection of plants existing in a near vacuum. However, recall that over the period of a year in the entomologist Frank Lutz's garden, he found at least 1500 species of insects when he took the time to look for them. Many of these insects were plant feeders, and as we saw in chapter 6, plant-feeding insects interact in many different ways with plants. But a great many of the insects that Lutz found were predators and parasites, as well, which were preying on the plant-feeding insects and each other. My humble guess is that, with a few common exceptions, you will not have seen a handful of these insect interactions in your garden even though there may be hundreds of individuals of any one species at any one time. These species and their thousands of interactions are what could be called the neutral insects: the little bugs that are present but never seen, the ones that run the world.

Does the concept of neutrality mean that these thousands of insects have no function in your garden? Not in the least. The insects and their actions and reactions are the garden's backbone. As many insects quietly go about their business of clearing broken bits of plants from the soil's surface, decomposing dead material, incorporating organic matter into the soil, dispersing animal droppings, gathering nectar, pollinating plants, chewing on leaves and flowers, and serving as dinner for birds and bats, another hoard of insects—the predators and parasitoids—are doing their best to find every last one and kill them. Such is life.

As we saw in chapter 7, the many interactions between insects fulfill the daily routines of nature—routines that we gardeners ignore completely in our blissful ignorance of how an ecosystem (of which a garden is but one artificial example) functions. By ignoring these insects and their millions of microscopic interactions, we banish the everyday processes of life, by which we all survive, to the abstraction of neutrality.

A simple example will suffice. We basically concede that ants are annoying little creatures that enter our homes and create problems we could live without. But there are likely to be tens of thousands of ants, representing several species, in your garden. You may rarely see an ant, yet they are there, and for every single ant you do see, there will be hundreds more underground. There is the old adage that "What you don't know won't hurt you." If you did know the ants were there, just under the surface, you would probably want to kill every last one of them because that is the way we humans often behave: without the benefit of forethought.

To many of us, I fear, the garden is simply a setting in which plants grow in a somewhat artfully arranged and cultivated manner. We have no sense that the garden is a microcosm of the Earth itself, with all the basic principles of ecology set in motion. We are virtually bereft of any connections between us, our garden, and our biological world. In reality, it is our inability to see the tremendous number of minor interactions in the garden that makes them detached from our own way of thinking. And it is our detachment that makes these unseen interactions neutral in the sense of having no connection with us.

If we have any inkling at all about biological interrelationships in our garden, they are queasy notions, tinged with a subliminal sense of terror concerning only a very basic thought: the arrival of that one dreaded insect that is going to decimate our plants and presumably our entire garden. We fear most the arrival of *the pest*.

Ants are ubiquitous. There are nearly 9000 described species in the world, so no matter where we garden, there will be ants. The ants shown here are dismembering a moth larva, which they will take back to their underground nest. Ants are one of the great recyclers of nutrients in our gardens.

Warfare, or Moments of Aggression

We gardeners know insects when we see them, and we see them most often when they attack our plants. At that point we feel compelled to engage in an all-fronts battle for victory over the foreign invaders. We attack nature at its own borders with our chosen weapons. The enlightened warriors will use trapping devices, such as tapes and sticky boards, or exclusionary defenses, such as spunbonded fibers and plant collars. The more aggressive types will head straight away to the arsenals, sloshing buckets of chemicals in every direction, slopping for all they're worth in the hope of killing every living thing. In their view, only then will all be quiet on the western front.

The concept of pest is entirely a human notion. There are no pests in nature. For humans, pests get to be pests when they teeter out of balance with their environment and do things we don't want them to do. Insect pests are the garden's flying weeds. Here's an example. A single aphid begins feeding on your rose stems. You most likely will not see a single aphid, or two, or ten. If a ladybird beetle larva feeds on these aphids you may never even know they were on your roses, and you will not even care. These aphids are not pests. If the population soars to, say, twenty aphids, you might see them for the first time; at this point they might suddenly become pests. The question becomes one of tolerance and how much of it you should have. In agricultural entomology (including nursery crops) there is something known as the economic threshold, or the point at which it becomes more economical to do something to save the crop than it is to do nothing and hope the problem will solve itself (which is entirely possible and economical). The economic threshold is hit when the grower goes into decisive action to save the plants.

In our gardens, this point is not so much economic as it is aesthetic or even worrisome. It arrives when a gardener fears for her plants and when she accepts the cost (in terms of time or supplies) of doing something or of gritting her teeth and doing nothing (some would say letting nature take its course). This decision becomes one of choice, not really necessity unless the gardener is highly devoted to the plant(s) under attack. It is at this point that a minimum amount of knowledge really can be of great benefit.

Gardeners tend to be fairly level-headed folks, but when insects threaten, the garden gloves come off and the boxing gloves go on. It is a gut reaction. No one wants to see their tomato plants reduced to piles of nutrient-rich frass by tomato worms, the edges of their prize-winning rose leaves cut into notches by leafcutter bees, or their dwarf mugo pine reduced to a skeleton by sawfly caterpillars. When these sorts of insects come to call, as often as not the gardener will first seek out the quick cure (whether mechanical or chemical) without thinking too much about the consequences of the action. This is a self-taught, cultural response, which is basically out of sync with the way a garden ought to work and all of nature, in fact, tries to work. We don't respond the same way, for example, when the thousands of other neutral interactions take place in the garden, the ones we rarely see or care about.

I couch garden neutrality in terms of a state of mind, that is, a neutrality of knowledge. Neutrality is essentially what we gardeners do not know and therefore do not fear; it's a sort of don't-ask-don't-tell, live-and-let-live, ignorance-is-bliss proposition. In contrast, warfare entails what we gardeners think we know and therefore think we fear: kill the invader and all else be damned. In reality, the garden is in a constant and agitated state of war, but we rarely see the results because we are concentrating only on the isolated big battle we choose to stand and fight. Once we learn that the garden is an island of insects engaged in perpetual warfare at all times, the idea of neutrality becomes simply the acceptance of the mundane interactions of insects with plants and each other. In other words, nature's way of doing things is typically brutal and effective. That's all right as long as we don't see it or think about it.

Unfortunately, when we haul out the pesticides we not only must face the battle, we usually end up prolonging the war we meant to end. By applying artificial agents such as chemicals to control plant-feeding insects, we create a vicious cycle of the worst kind. Being at the base of the food pyramid, a plant feeder is generally more numerous than the predators and parasitoids that eat it (think of the hordes of aphids and the relatively few ladybugs we see eating them). Because the plant feeder is numerically dominant to start with and is almost never entirely eliminated by insecticides, the pest rebounds from chemical attack much more readily, but now with few enemies. We must again use chemicals to control the dominant insect, but this time it rebounds a bit stronger than before it became genetically more resistant. This is the principle behind every agroecosystem that humans developed in the first half of the twentieth century, and it is our legacy to the planet: a cycle of chemical dependency.

"But," I hear you ask, "is there no cure for this dependency? No rehabilitation center to help us off this chemical treadmill?"

And I, the ever-careful biologist, can answer you confidently, "Well, yes. Of course there is! Sort of." The answer lies in yet another human notion, one that is now highly acceptable, if not all that well adopted, that insect champions might come along with our encouragement and rescue us (basically from ourselves).

Champions, or the Peace of Beneficial Insects

In *Silent Spring*, Rachel Carson first challenged us to think in broader terms than our basic gut reactions coupled to a trigger-finger on the plunger of a flitgun. She showed us, as Garrett "we-can-never-do-merely-one-thing" Hardin later suggested in his book *Living within Limits*, that in attempting to solve one insect problem, we could create many more problems than we could even imagine. The obvious example that comes to mind is DDT, which was first the wonder elixir and then the infiltrator of birds' eggs, fishes' brains, and mothers' milk. As a result of Carson's writing, we began to think about the connections between what we sowed and what we reaped, and we came to the conclusion that we might be ingesting and breathing a lot of stuff that we would be better off without. One of the positive outcomes of this thinking was that we realized it might be possible to set an insect to catch an insect. Thus was born, in the minds of many nonscientists, the idea that if you threw biological organisms at other biological organisms you would solve all your problems and also not upset the environment.

This notion of setting one insect, the beneficial one, to catch another, the pest, is what economic entomologists call biological control (biocontrol, for short). In fact, this is a very old concept, dating back hundreds of years to times when ant colonies were placed in citrus trees. Rope bridges were hung between trees to encourage lateral movement from tree to tree, and the trunk of each tree was surrounded by a shallow, water-filled moat to prevent the ants from running away. These ants would then keep other insects off the

Green lacewing larvae are voracious aphid predators and some of the best insect predators money can buy. Because they are sold as eggs and the emerging larvae do not move very far, green lacewings stay where you put them and do the job they're paid to do.

trees. Since the late 1800s, we humans have successfully used biological methods to control insects as well as introduced, noxious weeds. A number of books detail these efforts, such as those by DeBach (1970, 1974), Graham (1984), DeBach and Rosen (1991), and Flint and Dreistadt (1999). However, biological control agents have been established mostly for agricultural crops and rangeland weeds that grow on a large scale. It has only been within the last decade or

Ladybird beetle adults bought in a box are not a good value for your money. Their natural tendency is to "fly away home," even when aphids are abundant. In undisturbed systems ladybird beetles help maintain an equilibrium. However, when we speak of insect control in our gardens, we are admitting that populations have gone out of balance.

so that many gardeners have become involved directly with introducing insects into our small-scale gardens.

Lately we gardeners have been inundated with catalogs that sell biocontrol agents. On the surface, the idea is commendable. Realistically, however, and in my humble opinion, you would be much better off reading the next several chapters than buying bugs. This will have the net effect of saving money so that you can buy more plants. More plants are what will bring more beneficial insects to your garden. But I am getting ahead of myself.

Biological control, although often a wonderful, workable, and worthwhile approach, is not a panacea for all things gone wrong. As

Ladybird beetle larvae, like lacewing larvae, do not spread very far from their feeding grounds, but no one sells the larval stage. There would be no need to buy "good" insects as hedges against "bad" insects if we encouraged them all to live in our gardens.

I noted above, it works best on a large scale with crop plants and noxious weeds. The fact is, even in agroecosystems, when we try to correct the balance of nature by using beneficial insects, the technique only works successfully some of the time. Sometimes it doesn't work at all. If you look at the approach realistically, applying beneficial insects from a box is scarcely different than applying chemicals from a bottle. Using biological controls certainly makes us feel good, but the need to use either control method is the result of a basic imbalance in the garden. It is far better to have all the insects engaged in their routine natural battles than to search for a *cure,* which can be found in neither a bottle nor a box. To maintain a balanced garden is what we might call a return to the natural condition (or at least as natural as we can be in the artificial environment of a garden).

The Natural Condition

Under natural conditions, a number of biological and environmental factors exist that keep populations from increasing to a point where they become a nuisance to themselves. Another way to put this is that under unnatural conditions, populations can build up to a point where they tend to harm not only themselves but the environment on which they depend. (Some writers, such as Thomas Malthus and Garrett Hardin, have said that the human species is no exception to this basic ecological principle and that eventually we will discover these truths for ourselves.)

Populations are controlled by abiotic (environmental) agents such as floods, droughts, and freezing or by biotic (biological) factors such as predators, diseases, and competition for food, shelter, and mates. Classically, then, ecology has been the study of how all these factors interrelate to each other, how organisms are affected, and how they affect other organisms. The study of ecology is much like a puzzle for which we have few answers and, in many cases, most of the pieces of the puzzle are missing. Conceptually, it can be quite a challenge. For those so disposed, contemplating ecology can be a Zenlike experience.

Balance, or Harmony

Balance is what happens when nature is in harmony with itself. By harmony, I do not mean peace—there is no peace in nature. For nature to be in harmony all factions must be in constant battle. Darwin's followers call it "survival of the fittest." For instance, a plant seed must fall in the right place and on the right soil type to successfully germinate. Then it must outgrow any other seed of its own kind, or a different kind, if it is to survive. Then the plant must be able to endure anything that tries to eat it, stomp on it, burn it to the ground, freeze it, or desiccate it. In any one location, there might be tens or hundreds or thousands of individuals of the same species attempting to survive long enough to do one thing, procreate.

Then, too, there may be tens or hundreds or thousands of other plant species also trying to survive in the same place. This is what you see when you look at a meadow, a forest, or even a desert (especially after a rain). The grand and gloriously beautiful scene is the result of millions upon millions of tiny interactions. What you perceive to be an immutable scene is a constant unseen struggle, progressing second by second, minute by minute, hour by hour, and not just between plants and soil, plants and light, plants and water, but between plants and insects, plants and deer, plants and rabbits. Add to this the untold millions of interactions between insect and insect, insect and bird, insect and bat, and you have created a palette of intricate interdependencies that would stifle a computer the size of all outdoors.

When you look out over that beautiful mountain meadow, the one with millions of microscopic, interdependent actions, you are observing a balance or harmony that you accept as natural. This balance is what ecologists call a plant community (actually a meadow may represent many plant communities, but for the sake of simplicity, let's just call the whole thing a meadow community). These systems—the meadow, the forest, the desert—all represent what I have called the neutrality of the garden. Nothing is drawing attention to itself, everything is in its place and functioning normally.

Now imagine that you collect the seed of one lovely plant (gardeners always find plants they just must have), take it home, germinate it, pot up a healthy young plant, and eventually place it in your garden. You now have a single plant that is, in a manner of speaking, completely out of its natural element. All its normal elements are missing: soil, air, moisture, root microbes, insects (for instance, pollinators), competitors, and so on. It is a wonder that such a plant would grow at all, and any honest gardener will tell you that they've lost 99 percent of such plants. It is often difficult to take plants from their native environments and make them adapt to an exotic environment. Thus, it is generally much easier to grow plants that are native to the area where your garden exists than from far distant and exotic shores.

The mantra of the native-plant people is, naturally, to use native plants, the ones that are already in balance with the place you live. These are the plants that are most likely to do well under your growing conditions, to grow without the input of artificial watering or fertilizing, to be insect and pathogen resistant, and not to become noxious weeds. Although most of these notions are broadly true, they cannot be taken without certain caveats. Many times native plants are even more temperamental or touchy than introduced ones. I've never been able to grow native amelachiers, for example, because they die back, apparently from some fungus in my soil. On the other hand, our native enchanter's nightshade, which I encouraged one year, became an insidiously spreading menace the second. No one wants an obnoxious plant, no matter where it comes from. And the evidence that native plants are better adapted to local growing conditions is debatable when you consider that any of the hundreds of introduced plants such as Japanese honeysuckle, tree of heaven, English ivy, and kudzu grow so much better than natives that they wipe them out. In other words, many plants (especially those that become noxious weeds) are better adapted to growing in an environment (any environment) than the native plants they overrun.

I must turn aside, at this point, before I am ostracized by the native-plant people. Please do not take what I say the wrong way. I have nothing against native plants. As a biologist, I have spent years wandering among plants and animals. I love to see them growing in their native environments. In the past, I spent more time roaming the wilds than weeding my garden. I am as appalled by the release of invasive plants into our native environments as anyone, and the complete destruction that invasive plants cause alarms me. I merely state that it is possible, in the artifice of the garden, that introduced plants will work as well as native plants and usually with more pleasing results. The only thing that native plants and their environment have that most gardens don't is an inherent sense of place, a sense of balance.

David Cann (1998) perhaps described the notion of balance most succinctly when he said "Natural communities thrive because they represent *a structured and balanced assemblage of species* that change over time" (emphasis mine). Cann was reviewing the book *The Self-Sustaining Garden* by Peter Thompson, in which the notion of matching plants to the ecological conditions of your garden takes precedence over whether the plant is native. In Thompson's mind (never mind the specific plant's), the ultimate goal is to create a matrix of "roots, stems, foliage and flowers, which provides protection for its members and resists invasion by outsiders." In other words, the object is to create an ecological community that mimics how a real or natural plant community acts—the way a forest or a meadow behaves, for instance. As Thompson says, "Any plant can be planted in combination with others with the intention of forming a self-sustaining community."

This sense of balance, of community, of basic naturalness has been espoused by several gardener writers of late, and I believe it will become increasingly a theme in the new millennium. This theme is perhaps best expressed in the title of Mirabel Osler's book *A Gentle Plea for Chaos*. Let us, Osler says, have a little "amiable disorder." And others agree. Sara Stein in *Noah's Garden*, Janet Marinelli in *Stalking the Wild Amaranth*, and Evan Eisenberg in *The Ecology of Eden* entreat us to adopt the naturalistic approach to gardening. Not necessarily the natural way, but in a way like nature. Not surprisingly, they endorse the concept of the interdependency of organisms, thus allowing all creatures a share of the garden. These authors seem to suggest that the garden needs a whole lot more diversity and a lot less stuffiness. Perhaps they have not gone so far as to invite in all the insects, but that is what they mean to say. I say, "What's good for the garden is good for the bugs."

9

Diversity:
The Basis of a Balanced Garden

If you desire a balanced garden, one that maintains an even keel with no great upswings in aphid populations this week, scale populations the next, or bean beetles a month from now, then you should be keenly interested in what promotes balance in the garden. In the last chapter, I noted that insect outbreaks, which are a prime example of instability, are a result of a disruption in the inherent balance of organisms—what some would call the natural order of things. (The concept of balance applies as much to weeds or fungi as it does to insects, but I will continue to use insects as the basis for the discussions to follow.)

In this chapter, I discuss the primary ways in which plant diversity is envisioned in nature: a theoretical lesson, if you will, in how nature builds a plant and a plant community so that it can provide, in turn, a complex universe for insects to inhabit. We will explore what gardeners must *see* to plan their gardens to include diversity at all levels. In chapter 10, we will then explore what gardeners can *do* to improve the diversity of their gardens. Seeing and doing are two completely different approaches to gardening.

Diversity in Principle

I have been told by an ecologist, whose opinion I respect highly, that the term *stability* has a precise meaning in the field of ecology: the ability of a system to return to equilibrium following a disturbance. When I use the term *stability* below, I mean retaining an equilibrium, a balance, a stable *state*, rather than restoring balance following some sort of an event. With this disclaimer, and perhaps a bit of license, the discussions that follow are designed to explain how diversity and stability, or balance, are intertwined.

Ecologists have convincingly documented that stable ecological systems (ecosystems) also have a great degree of biological diversity. What this says, in simple terms, is that a system that has lots of different plant and animal species also has a high degree of stability, or balance. This point might be envisioned as a three-tiered system: biological *diversity* increases biological *complexity*, which increases biological *stability* (or balance). In terms of the garden, we can say that the more biologically diverse the garden, the more complex it is, and therefore the more stable it becomes. Alternatively, the less biologically diverse the garden, the less complex it is, and the more easily it becomes unstable (that is, unbalanced). In this second example, the garden has no structural framework, it begins to wobble at the slightest provocation, and when we least expect it, the garden metaphorically falls over.

This last concept is exactly what happens in an agricultural system. If you plant 1000 acres of corn, you will have a very simple ecosystem, a monoculture. This big, mouthwatering patch of corn will be screaming to all its detractors "There's a free meal here. Come and get it." And what will stop insects such as the voracious corn earworm from chomping down? Well, nothing. Nothing, that is, except a ready dose of insecticides or maybe a breeding program for resistant plant varieties. With the insecticides and resistant plants arrives a further numbing of the agroecosystem because plant-feeding pest species quickly become resistant to insecticides and overcome the resistance of engineered plants. Essentially, insects beat us at our own game, which ensures a never-ending battle for control and an ever-worsening cycle into total chemical dependency.

Now consider your own garden. How close to a monoculture is it? For one thing, you have a lawn that is essentially a dead, flat space. It is a monoculture in the sense that it has scarcely any variety (and dandelions don't count for much), and it is drastically disturbed on a routine basis. What could live there? Then there is the rose garden. Talk about a chemically dependent monoculture! You've got a dozen insect problems, half a dozen fungus problems, mite problems, bacterial problems, and the sorts of problems we gardeners do not talk about in polite society. The vegetable garden is likely to be a pristine, hoed, rowed, barren sort of place. The annuals, hybridized for looks and no nectar and pollen, have scant appeal to any living, nonhuman being. They are bedded out like corn and are in the ground for only a few months, presenting a scorched-earth policy for half a year during late fall, winter, and early spring. Perennials might be a source of domestic bliss for a few bugs now and again if they weren't deadheaded, cleaned up every few weeks, and scraped off the earth before winter sets in. The shrubs might offer some free housing units for a few odd insects, but generally there are scarcely more than three yews and a juniper in any given garden.

If you looked at your own garden from the point of view of where organisms might live—and let's assume for a moment that you want them to live—would your picture be as bleak as the one I've just drawn? Do you see, in some respects, what I mean by simplicity, absence of diversity, absence of complexity? I hope so, because it is this absence that is what opens your garden to the susceptibility of outbreaks (instability) of the insects that you do not want to be there. These are what we want to call pests, but are, in fact, only insects that have gone out of balance with their environment.

It is perhaps easiest to illustrate a basic ecological point using the model of artificial biological control. In biological control, it is essentially impossible for a parasite or a predator to find every last individual of what it is hoping to eat. If a parasite finds the very last aphid in your rose garden, for example, what will its emerging children eat? They will have to fly away to someone else's rose garden, and then you must hope they eventually find their back to yours because the aphids most surely will. There is nothing you can do to permanently eliminate aphids from your garden unless you undertake constant vigil coupled with a bottle of noxious poison.

As we saw in chapter 8, it is an ecological fact that plant feeders are more numerous than the parasites or predators that eat them. Because aphids are more reproductively gifted than their dull-witted adversaries, they will invariably outbreed them. In a balanced system, the parasites will always be within striking distance of killing the last few dozen aphids, but they will never quite make it. And this is a good thing. You want those parasites in your garden constantly looking for that last aphid and never finding it, because that is why a natural system is balanced. In nature, neither parasites nor predators will ever completely win the battle because as soon as they do they are dead, so are their progeny, and so is the species.

In the garden, what we are looking for is not artificial biological control as purchased from a catalog and applied from a box, but natural control as generated from the environment, from the garden itself. Theoretically, this should be easy enough to accomplish, just let nature do its thing. But we gardeners (or commercial farmers, for that matter) are not so good at letting nature have its way. Staying with the aphid analogy for a moment, for natural control to work in your garden there must be a residual population of aphids somewhere, perhaps tucked away where you cannot see it, and some of these aphids must have a few parasites in them. It is this little stash of aphids and their natural enemies that keeps your garden in balance. The more the aphids reproduce the more the parasites reproduce. If the parasites must fly in from someone else's garden (someone who obviously appreciates them more than you do), then your aphids are going to have a commanding head start at producing more young than their enemies. The aphids will win the battle, and out will come the insecticides. To keep the aphid from becoming a pest, you need to have both the aphid and the parasite close together in your garden. The psychological hurdle is that the gardener must endure a certain degree of detachment from this constant ballet of up and down cycles. When these cycles first begin to develop, it is not a good idea to break them, if you can help it.

These ants are tending aphids from which they
obtain honeydew, a sugary, excretory substance. The ants protect their perpetual food
source, at least in theory, by chasing away any
other insect that comes near, but they are not
necessarily successful. Near the bottom of the
stem, a minute parasitic wasp can be seen laying an egg in one of the aphids. On the upper
left of the stem, a golden aphid stands out
from among the blackish green ones. This
aphid has already been parasitized and is dead.
Half the aphids on the stem could be parasitized and neither the ant nor the gardener
would know it until they turn brown. Parasitic
wasps have a knack for finding their hosts,
even in a greenhouse.

A crab spider has lost its advantage
of surprise as a predator when it
wanders onto a background of grape
hyacinth that exposes it as potential
danger to passing insects.

When camouflaged on a flower of similar color, the crab spider is more likely to have a chance of capturing its prey, in this case an adult sawfly. Sawfly larvae normally eat plants, but the adult acts as a pollinator, thus increasing the seed set on California poppy. By human standards, the sawfly is beneficial at one life stage, yet harmful at another. Is the spider beneficial when it kills the adult so she cannot produce plant-eating larvae or harmful when it kills the adult pollinator? Some questions are best left for nature to answer.

Here the same well-camouflaged crab spider has killed a honey bee. Some consider this beneficial: honey bees are an exotic, nonnative species, after all, and they may rarely even kill people. Others would disagree, of course.

The concept of host and parasite (as well as host and predator) is a simple ecological principle. Having recounted this simple parable of the aphid and the parasite, we need to ratchet the discussion up a notch in complexity. The next step is to think of the concept in terms of an aphid species in your garden being stalked and hunted by flower fly larvae, ladybird beetle larvae and adults, lacewing larvae and adults, and half a dozen different species of parasites. Then, add to this the parasites that specialize in attacking only flower fly or lacewing larvae or the ones that attack ladybird beetle larvae. There are parasites that attack only ladybug adults. Then there are the hyperparasites (hyperparasitoids) that attack only other parasites that attack aphids, ladybug adults, or other insects. Soon it becomes apparent that around any aphid species there is a whole community of other insects that depend on that aphid. When you think about spraying that infestation of aphids on your roses, remember this chain of organisms—many working to control the aphids and many working to control each other. It is impossible for a gardener to orchestrate this complex set of interactions, for example, selecting the ladybird beetles while eliminating their parasites. The best thing to do is accept the whole community and let all these species do the battling among themselves. No single insect species will win, and that is exactly what you want in your garden.

So far, we have looked at only a single aphid species and the picture becomes very complex indeed. But there are other aphid species. Many plant-feeding aphids are generalists that feed on different genera and families of plants. But some aphid species may be rather specific. For example, there are species that feed only on members of the onion family and there are others that feed only on members of the milkweed family. Often, the more host specific the aphid feeder, the more aphid specific are the insects that attack them. In the typical garden, then, you will have generalist and specialist plant feeders, generalist and specialist predators, and generalist and specialist parasites. The complexity is ratcheted up yet again.

But hopefully you have more than aphids in your garden. There are moth larvae and gall formers, beetle grubs and leaf miners, fly maggots and stem borers, bee larvae and leaf chewers, and on and on. Each of these insects, interacting with its specific or general plant hosts, has communities of other insect predators and parasites keeping its numbers in balance. And so it goes, interaction piling upon interaction, species piling upon species, until you have an ecosystem of insects and plants and their environment exactly like the plant matrix that Peter Thompson describes in his book *The Self-Sustaining Garden.* This matrix of behavioral interactions is what keeps natural systems in balance.

The behavioral matrix I have been describing is a highly complex set of interactions based on the ecologist's mantra: biological diversity increases biological complexity, which increases biological stability (balance). Now that we have examined diversity from a basic ecological viewpoint, we have arrived at one of the great powers that drives an ecosystem. It is this power that I am asking you to invite into your garden, or at least not to turn away. Consciously you could not design these webs of life even if you tried, but in my opinion you can develop the garden and plant structure that will ensure the same effect. If you take a few small steps, most of these interactions will come to you. As a practical primer to achieving diversity, let's next turn our attention to the application of the basic principles of diversity.

Plant Diversity

As we saw in chapter 6, there are endless ways in which insects interact with plants, from roots to seeds, from bulbs to stems, from underground to underwater. Even a single resource such as leaves, for example, have their specialist insects that mine them, gall them, cut circles out of them, roll them into tubes, tie them together, or chew them up and make fungus. These plant-feeding insects, in turn, provide the basis for thousands of other insect interactions with generalist predators, such as praying mantids, to wasps that hunt only caterpillars to parasites that specialize in the eggs of one specific kind of insect. Whatever the boundless interactions of insects with each other might entail, the basis for their biological di-

A spider-hunting wasp has paralyzed its prey and is dragging it off to place in an underground cell, where the wasp's larva will eventually consume the spider. Spiders can appear to be beneficial or harmful depending on what they eat. What seems like a simple act of killing for the spider-hunting wasp is food for complex human thought.

versity is the humble plant. Thus, we must begin at the level of the single plant.

As I demonstrated with the aphid and its parasite, I am going to build on the diversity of a single plant from the ground up. Even a single species of plant provides hundreds of levels of diversity. First, a plant has parts, lots of parts. It has a seed that can be eaten by birds, a seedling that can be eaten by cutworms, a stem that can be mined by moth larvae, a leaf that can be sliced and diced by leafcutter bees, a flower that provides nectar to butterflies, pollen that is taken by bees, a seedhead that can be galled by fruit flies, and eventually a corpse that can be digested by beetles.

Second, a plant has an architectural element that creates diversity. This architecture lies not just in various levels of height and spread, but in the way the plant branches, the way the leaves are shaped, the way the flower is shaped, how the nectar and pollen are produced and displayed, and even how the plant dies—slowly, in bits and pieces over many years, such as a tree, or in one season, such as an annual.

Third, a plant provides multiple levels of diversity in its phenology, that is, its development over time. The plant's development unfolds within a framework that allows endless permutations of interaction with insects over minutes, hours, days, weeks, months, and years (in the case of long-lived plants). Taken together, a single plant has an enormous potential to provide what an insect needs, namely food, shelter, and the opportunity to mate.

Now proceed to the next step. Add one more plant of the same species to the first. It is well known that two objects cannot occupy the same space at the same time. Likewise, two plants cannot occupy exactly the same space. Even if they are planted an inch apart and are genetically identical, the plants will begin to behave differently from each other. One plant will affect the other by causing it to grow more slowly, forcing it to bend toward the light, or stealing more moisture or more sun. Thus, even two plants of the same species growing at the same time in nearly the same space with identical soil, temperature, humidity, and moisture present more diversity to an insect than a single plant alone. One insect species may be attracted to the plant that is slightly stunted or stressed from the competition of its stronger neighbor, whereas a different insect species might be attracted to the vigor of the stronger plant. Insects are especially clever at locating plants that fit their precise needs.

The lesson learned from even the simplest case of plant diversity, with two individuals of the same species, is that different organisms sort themselves out to fit different biological conditions. Just as two plants will attempt to adapt themselves to the situation they are in, so will insects sort themselves to the plants that are available. In many (if not most) cases, garden plants of the same species will not be genetically identical (only if cloned will they be truly the same). Thus, a single species of plant offers much more genetic variation than a single plant. One species of plant, then, can present diversity over time based on genetic and environmental factors. If two plants can provide more diversity than a single plant, then the more plants you grow and the more different kinds of plants you grow, the greater will become the structure, complexity, and diversity of the insect community using the plants as their home.

Obviously, a single plant species provides a fairly simple example. With a simple example you get fairly simple results, but even so, it is a great deal more complex than agricultural monoculture. In agriculture, a single type of plant is grown in controlled rows at specific distances in chemically altered soil that is weeded with discs and chemicals. The result is that not even two plants can interact very much. There is little or no diversity in such a system. Also, there are hundreds or thousands of acres under such constant supervision and under a constant barrage of insecticides, miticides, fungicides, and herbicides. Basically, the only organisms that have a chance to survive under such circumstances are the plant feeders, and even they are so overwhelmed as to be virtually nonexistent. The lesson that might be learned here is that within a well-structured, simplistic, monolithic patch of plants, there is no natural or ecological balance at work. The overlord of such a plant prison—for that is what a monoculture truly is, little getting in or out—must be on constant chemical guard to keep it that way.

A large variety of genetic diversity, including different varieties of the same plants and different species of plants, will increase the diversity of plant parts, architectures, and developmental sequences, which in turn will increase the complexity of the garden. But additional steps may be implemented to increase and enhance the complexity of these plants by simply increasing the number of habitats within the garden.

Habitat Diversity

Plants are influenced by the habitat in which they grow. As gardeners, we generally know that some plants grow in soil, others underwater, free-floating on top of water, attached to other plants (for example, vines, epiphytic orchids, bromeliads), and so on. We know that plants grow in sun or shade; some plants that grow in sun prefer sand, whereas others prefer humus-enriched soils. We know that some plants grow on gravelly limestone outcrops, others on acidic organic humus. We know that alpine plants and cloud forests grow at high elevations, whereas rain forest plants grow in steamy, lowland jungles. We know that, unless mowed, fields of grass are soon invaded by trees that shade out the leaves and outcompete the roots for moisture. We know that some forests must burn down periodically to restore their youthful vigor. As gardeners, we may know a lot of these things—and thus be closet ecologists. Yet, a lot of us pay no attention to this knowledge when we plan our gardens.

Something as basic as soil, for example, may allow us to divide the garden into subsections, thus doubling or tripling the variety of plants we grow. And every increase in plant diversity provides a potential increase in insect diversity, which is another step toward a stable garden. We might have greasy, clay soil in one area of the garden and soft, spongy humus in another. We could have a wet, boggy spot and a dry, hardpan spot. We should first adapt ourselves to these conditions and try to find the correct plant to place in the correct spot. Often, however, we end up trying to standardize our garden so it will be the same all over. Most often we want to lay down a constructed plan we have in mind, when we should be using the soil and the site as the bases for the plan. We drain and raise the wet spots so they will be as high, level, and dry as the remainder of the garden. We add gypsum and humus and sand to the clay bands to make them fluffy and light like the humus-filled spots. Eventually, we end up with a soil base that attempts to be all things to all plants. But we have also simplified the habitat diversity with which we were naturally endowed.

Full sun, full shade, and the complex variations of the two provide endless possibilities for plant combinations, all of which will increase diversity. Although a natural meadow appears to be in full sunlight, many of the smaller plants lie in shadows cast by taller or more robust species. Over time, plant species sort themselves out according to their light tolerances, and insects will sort themselves out based on their plant preferences.

Habitats with water features such as lakes, ponds, streams, seeps, and springs will each have a certain set of plant species adapted to life under specific conditions, thus increasing the total number of different plant (and insect) species associated with these habitats. A single pond may have wide, shallow, boggy areas that supply special requirements to one set of plants and submerged mud banks varying from an inch to several feet that supply requirements for several dozen more species. Some of the differences within narrow bands of submerged soil may provide microhabitats in which a few species of plants will grow and no others. If you consider the potential habitat diversity in a single pond, you will see that even this simple structure can have a tremendous subset of habitats that increase the diversity many times over what the pond appears capable of sustaining.

The foregoing comments are not to imply that every plant species must have completely restricted and precise growing conditions. If this were true, the variety of species growing in our gardens would be considerably reduced. Some plant species are extremely adaptable to a wide variety of wet or dry conditions, for instance, and thus present a rather unpredictable picture of what requirements they truly need. The bald cypress, for example, which is always associated with ponds and shallow water, actually grows better in an "unnatural" habitat than in water-logged soil. The tallest trees I have seen are growing in hard-packed gravel and clay in front of the National Museum of Natural History, where I work. Their compacted roots have been trodden upon by hundreds of thousands of tourists' shoes. Yet, the trees still thrive. As it turns out, bald cypresses actually prefer dry soil if given no competition from other plants, but they have adapted to live in wet soil, where they outcompete other tree species.

In discussing habitats, we should remember that many plants are more adaptable than we realize. It is true that some plants grow in the deepest shade or the brightest sun, others grow only in shallow water or in the driest of sands. Although there are truly specialized species, many plants have generalized requirements and remain somewhat adaptable to varying environmental conditions. Two secrets of being a good gardener are the ability to ignore what we think we know about plants' requirements and the willingness to kill a large number of plants in our quest to discover what we can get away with. Sometimes, as in the case of the bald cypress, we can fool Mother Nature entirely—or at least we can fool ourselves into believing that we know what a plant wants, when we haven't a real clue at all.

Habitat Stability

Habitat diversity appears to be the cause of increased species richness and thus stability, however, some ecologists suggest that, in some cases, the situation may be just the reverse. That is, habitats with great stability allow populations of many different organisms to build up in them—to become more complex. To gain some understanding of this concept, all a gardener needs to envision is a habitat such as the lawn, which is whacked back every week or so. Few organisms can live under such volatile conditions, and few species will have the opportunity to build up. Every week most insects that are living in the lawn will be ground to bits, and the invasion process will have to start over again. If this were a stable meadow, diverse insect species would have a chance to accumulate.

Although the concept of habitat stability is a fair and interesting debate among theoretical ecologists, the normal gardener need not lose any sleep over the finer details. If you have a huge lawn surrounded by annuals, then chances are you will not be able to establish a garden lush with biological diversity. Basically, it cannot be done—regardless of which direction you run the theory. So it's not worth worrying about which came first, the stability or the diversity.

If you are building a new garden from the ground up, so to speak, then you will most likely be increasing the diversity of everything to begin with, so you've already begun the process. For most of us, this increase in plant and habitat diversity will promote places for plant feeders to eat and refuges for predators and parasitoids to hide, so that the two groups of insects will be better able to interact with each other when the need arises. If you have followed the ecological plot of the story so far, then you should be ready to tackle the next phase of what was initially an absurd notion: the more insects in your garden, the better off it will be. If you build a more complex garden, the insects will come and the garden will become more manageable, naturally.

Jay W. McRoberts, M.D., professional butterfly wrangler and owner and proprietor of Butterflies on the Potomac, Dickerson, Maryland, raises living stock for insect zoos, butterfly shows, and weddings. Few gardeners would emulate McRoberts, but we could be more hospitable when it comes to associating with insects and inviting them into our gardens.

PART III

Insects and Humans:
The Gardener's Perspective

Nature writes, gardeners edit.

ROGER SWAIN
Groundwork: A Gardener's Ecology, 1994

If you have been following the text up to this point, you may realize that insects are not really smarter than humans, it's just that we are vastly outnumbered by a group of organisms that has had about 400 million more years to adapt to the plant world. Gardeners will never have the time to catch up to insects, so we might as well make up for our comparative youth by examining how insects fit into the scheme of things (which we just did in part II) and then using this knowledge to our own advantage (which we are going to do in part III). If there is one thing that humans are good at, it is adapting almost everything to our advantage—which, unfortunately, is also one of our great downfalls.

As a species, we poor, misbegotten humans sometimes attempt to apply simple, quick solutions to solve complex, long-term problems (or, more correctly, perceived problems). Nowhere is this more evident than in our gardens, our small-scale, time-limited view of the Earth, where we not so long ago believed that all problems could be solved quickly by waving about the magic wand of chemical cleansers. Somewhere along our pathway of chemical simplicity, we forgot that the garden is a complex interaction of biological principles. Simplicity goes against all the grains in nature's hourglass. What we gardeners need is not simplicity—and its simplistic thinking—but a good dose of complexity. It is only through complexity that we can reestablish biological balance in our gardens, and only through balance that we can find peace from the constant battle we gardeners imagine we must face from unwanted, six-legged invaders. Ironically, by following a path of complex biological interactions within the garden, we gardeners could ultimately stop worrying about problems (as best we can) and begin appreciating the garden as a place of living balance.

In part III, I discuss positive steps that can be taken to create a garden and a mind-set of balance. It is not as if we are entirely helpless in this matter. As I will point out in chapters 10 and 11, there are many steps we can take to help restore the garden to a biologically more complex space. Through a process of plant and garden habitat diversification, we can create an environment that allows insects to undertake the intricate and tough balancing act, and thus relieve ourselves of the work involved in insect population regulation.

In addition to purely practical advice about planting the garden for insects, in chapter 12, I address some fears that gardeners may have about the mere presence of insects. With the exception of butterflies, we tend to fear almost all the insects we see. Fear is simply the product of unclear or misleading information about insects. I will do my best to explain some of this hysteria as well as to delve into the spiritual space of the vegetable or produce gardener—a space particularly sacrosanct in its devotion to the anti-insect mind-set. In chapter 13, I propose that, in fact, most of the historical interactions between humans and insects have been positive. The exceptions, however, seem to be what most people remember. Perhaps we have nearly reached a point where we might attempt to change our perceptions about insects and, in so doing, change ourselves as well. In chapter 14, I attempt to objectively and realistically summarize the gardener's relationship to insects—assuming that we wish to be realistic gardeners and stewards of the world we live in. If this is not the case, we might want to rethink our approach to gardening.

10

Increasing Diversity in the Garden

In this chapter, I discuss what is meant by diversity as it relates to both the plants and the physical structure of the habitat we call our garden. This information can help increase the number of insect species found in our gardens. As I have said, if we take care of the insects, the rest of the garden, with all its potential weeds and pathogens, will largely follow suit. But be forewarned, anyone who follows the suggestions given in this chapter will find their favorite plants taking on the dreaded mantle of weed when the garden shifts from muscle-numbing workhorse to natural-working wonder. You will be cursing those endless hosta and daylily seedlings, as I do, or you will be removing those native bloodroot and mayapple roots before they overrun the entire garden. Maybe you will be giving away heaps of hardy cyclamen and thyme, as I am forced to do occasionally, all to keep the garden from disappearing under a blanket created by its own boundless, competitive, and interactive exuberance.

All the components and complexity that create diversity make it sound as if you need a computer and a hundred years to plan out next year's garden. Such precision would, of course, result in total failure. Even with years of experience, you cannot possibly get all this natural stuff right the first, second, or even hundredth time. The biological world is an ever-changing universe that poses endless challenges to the gardener. A windstorm may topple a tree and change a shady garden to a sunny one. A hailstorm might shred a sunny garden to the ground. A disease might take out a favorite dogwood. A gopher might dig up an annual bed. But these are only the big things—the ones that the gardener sees. There are a billion little interactions going on in the garden all the time, and nary a single one is noticed. Bacteria and fungi work in the soil, plants and insects compete between and among each other, nutrients and trace elements are leached out of the soil and reenter it. Chemical, biochemical, and physical reactions of endless varieties abound. Against this incalculable chaos, no human mind can make an ounce of headway based on a reasoned approach to gardening. So why bother?

We usually bother so much that we end up creating the opposite effect of what we are striving to achieve. That is, by attempting to be in command of everything at all times—from soil fertility to insect and weed reduction—we tamper with forces over which we have little control. We gardeners often feel as if we should be doing something, and then feel better simply because we did it. Often we don't even know if the results were positive. The garden literature is resplendent with anecdotal affirmations of progress created by doing something, the evidence of which would be inadmissible in any court of science.

So if you really want to do something that matters in your garden, might I suggest something that matters in the long haul, that makes gardening easier and more enjoyable, and leaves the land and its resident plants and animals better off—something that will leave yourself, your children, and your grandchildren frolicking in the bargain? What I offer here is the basis for a sane attempt at gardening based on the principles of diversity as discussed in chapter 9. Certainly, if we know principles, we have a better chance to approach a natural balance of all elements—whether imagined to be good or evil—that combine to form our gardens. We shall put those principles into practice in two, essentially intertwined ways, namely increasing a garden's plant structure and its physical, or habitat, structure.

Sometimes a single plant can attract a diversity of insects. Although the butterfly bush is one plant well known for its attractive powers, the goldenrod shown here is equally attractive. Unjustly maligned by some, goldenrod is a fine late summer and fall plant for attracting insects. The cultivar shown here, *Solidago rugosa* 'Fireworks', is a semiweeping variety that splays out in all directions, the better to attract an assortment of insects.

A sulfur butterfly is content to find nectar within the goldenrod flowers. Its larvae do not feed on this plant, so the butterfly has no real attachment other than as a source of adult food.

This predatory eumenid wasp is chewing a small insect it found on goldenrod flowers that are past the nectar-giving stage. Insect larvae may live in the flower heads and continue to do so until the heads dry up and go to seed. These hidden insects can serve as food for others that visit the plant in search of protein, not nectar.

This adult flower fly is searching the goldenrod for nectar. If the plant stems were covered with aphids, the fly would also lay eggs and its larvae would feed on them. Thus, some plants serve multiple purposes for different stages of the same insect.

This katydid may have alighted on the goldenrod to eat its flowers—petals, nectar, pollen, and all. Possibly it planned to dine on leaves and slow-moving insects, or it may simply have been seeking cover.

Plant Structure

By plant structure, I mean the way plants are viewed in the present and over a period of time. As we saw in chapter 9, plants have physical structure (roots, bulbs, leaves, stems, flowers, for example), architectural structure (upright, weeping, trailing), and temporal structure (daily and seasonal expressions of both physical and architectural components). Each plant will have aspects of all these elements and will have a role to play at some time or some place in the garden. In the plant world, the crocus might be considered just right for a single, short, awakening pulse in the spring—a first call to the bees—whereas an ancient oak provides insects shelter and food for centuries. Both plants perform quite different functions in the ecosystem and the garden.

The ideal method of increasing overall plant structure is to weave the elements of physical, architectural, and temporal elements into an orchestral pattern that will fill up a garden's time and space for whatever seasonality it might have. (Not all gardens are created equal, remember, so we must allow that Californian and Alaskan gardens have different plant structure. The ecological elements will work with the same principles, but the players will be different in each area.) The simplest, most obvious way to accomplish this weaving is to grow plants that differ and to grow lots of them. Unfortunately, this simplistic approach to a complicated biological and ecological phenomenon is likely to be foreign to some gardeners. A lawn and a rose bed might simply be enough to take care of, you might think, or a lawn and a vegetable patch. If this is what a gardener wants to do, fine. I am not suggesting the total abandonment of all known gardening sense. But by underpinning the rose garden with a few low-growing plants, a gardener could add a minimal layer of diversity to the garden, thus improving the degree of complexity. Or if a gardener interplanted vegetables with some supposed attractant or repellent plants, she might achieve the same results, regardless of their purported powers.

This basic sort of plant combining (companion planting, or interplanting) has been suggested for years. Whether in the hippy sixties, the Mother Earth seventies, the New Age eighties, or the agriculturally sustainable nineties, it is a concerted attempt to get plants to grow together for the betterment of each other and of the local environment. Marigolds planted around roses, for example, are often touted as repellents for nematodes, as is planting wormwood to repel flea beetles from cabbage. Not to wade into the needlessly controversial, but the most likely effect of placing one plant to repel insects on another plant is the psychological comfort it gives to the gardener. There is every reason to believe that companion planting really works, but scarcely any chance at all that it works for the reasons imparted to its putatively repellent nature. If anything, it is the attractive nature of companion plants that saves the day, and the reason for this is increased biological diversity.

Companion plants offer two basic levels of protection to the garden. Perhaps the least obvious benefit is the confusion they provide to herbivores, which recognize their preferred host plant by sight and volatile chemicals produced by the plant. When many different plants are mixed together, a degree of confusion, or dilution, is created so that an herbivore cannot readily locate its host plant. In principle, the gardener is simply camouflaging vulnerable plants within the clutter of unsusceptible ones—the old shell game, so to speak. The second and more obvious benefit of companion plants is that they offer areas for shelter, food for adults, and opportunities to mate for predators and parasites. Of this there is no doubt. If only the target plants are grown, roses or cabbages, for example, the surrounding areas would be barren, at best. Because companion plants provide an area of protection conducive to survival, as opposed to a barren wasteland where no life can exist, they provide a reservoir in which insect interactions can routinely take place and from which predators and parasites emerge to feast on the insects we object to.

In truth, all plants are companion plants. There is no easier method to increase structure in a garden than adding plants, so let's look at the sorts of structure we should be introducing.

Physical and Architectural Structure

A single plant has physical elements such as leaves, roots, or stems. It also has an architectural structure that relates to size, shape, density, the number of different plant feeders it can support, and ultimately its effect in the garden. A single plant, however, does not a garden make. Neither a single beautiful bonsai nor a bed of roses offers much in the way of structure to a garden. A planting of 500 different kinds of daylilies or of 700 different cultivars of bearded irises, no matter how many acres in size, is not a garden: it is a cornfield without the corn, or at least the environmental equivalent of a cornfield. When we plant a half dozen irises and a similar number of daylilies together in our front garden we begin, ever so slightly, to increase its physical structure (not to mention the temporal structure, which we will come to shortly).

Combining the elements of plant structure creates diversity both biologically and visually. The differences between an iris and a daylily, when not in flower, however, are not really that great: both are knee-level bunches of leaves that point upward; both produce big, gaudy flowers that scarcely any tiny parasitic wasp or predator could love. Neither type of plant is capable of maintaining interest (either the gardener's or the insect's) for more than a few weeks each. Still, the two different plants taken together offer more than either plant alone.

If we continue adding the standard perennial assortment of plants to the mix—some taller or shorter, some fatter or thinner, some with bigger or smaller flowers—we eventually end up with an herbaceous border, which really just amounts to a cluster of companion plants. The multiple combinations and interactions of all these plants together offer more protection for insects than any single plant or multiples of the same plant could. But still we can do more.

Any border might range from groundcovers at 15 centimeters (6 inches) in height to colossal daisies 2 meters (6.4 feet) in height. To this, in the background somewhere or at one corner of the border, could be added an evergreen or two, varieties that don't grow terribly high. This would give shelter to birds. A deciduous shrub, say a butterfly bush, could be planted along a fence or even in the middle of a bed. This would attract passing butterflies. Bulbs, annuals, and even vegetables could be planted among the other border plants to provide food and shelter for roaming insects. In this manner, we build an artificial community of diverse plants that provide different options for life at varying times of the year. In effect, we attempt to create a garden plant community that mimics a natural plant community. (I will return to this concept and explain its significance at the end of the chapter.)

If this idea sounds rather simple, or much like a traditional garden, or that I've just reinvented the wheel, then take a look at most suburban plots of land and tell me what you see: lawns and foundation plantings, lawns and azaleas, lawns and yews, lawns and annual beds, lawns and roses. Bland. Dull. Boring. Many of these so-called gardens could use some reinvention, even if only to make the lives of passersby a little more enjoyable. Never mind the bugs.

We gardeners have at our disposal many plants that provide both architectural interest and physical shelter for a variety of organisms. These include all manner of herbaceous perennials, annuals, shrubs, trees, bulbs, grasses, and water plants that come in sizes from miniature to gigantic. They produce different-sized flowers, many of which appeal to special kinds of insects, but all of which, taken together, appeal to a variety of insects. If we add temporal elements to the physical, we increase the diversity of a garden's structure exponentially.

Temporal Structure

It is tempting to look upon the garden as a one-season wonder, and generally, if there is any single season that people are garden oriented, it is the spring. In some areas, especially low-lying deserts, spring is both a glorious time and a buggy time. That is because insects must compress their sexually active lives into a few short weeks to produce another generation. The new generation will find shelter before the oncoming heat wave and will either require a deep, season-long rest (aestivation) or confinement to evening and morning foraging. Insects that live in hot, dry conditions adjust their lives to produce offspring at the optimum times of the year, but they are present in some form—egg, larva, pupa, or adult—at all times of the year, whether the gardener can see them or not.

Life in higher-elevation deserts might involve two wet seasons, in spring and in late summer/early fall. Here the same insect species may have two procreative seasons; it is also possible that some insect species will have adapted to the spring season and other species to the late summer season. In this manner, the number of different kinds of insects might be increased in the same place, but these are active at different times of the year. This is a simple example of temporal diversity: different species at different times, yet all using the same space.

Of course, different geographic areas will experience different sorts of temporal diversity. In extremely cold climates, such as Alaska, the optimum time for maximum insect diversity would be midyear in a season that is shortened at both ends. Here, all the insects must survive extremely cold winters, which they are adapted to do without any help from their human neighbors. The point is that most insects survive through all manner of natural conditions in the areas they are adapted to live in, and they take advantage of good times to reproduce themselves.

One thing that gardeners can do, then, is to make certain that when the good times roll for insects, there is something in the garden for insects to roll into. This means that the more actively the garden is growing at any given point in time, the better the chance that a diversity of insects will stay around, reproduce, and be available to interact with each other in a balanced way. This is especially true of predators and parasitoids, because if they do not find prey or hosts they will leave an area in search of better hunting. The overall effect of this departure is that predator and parasitoid interactions become destabilized and outbreaks of plant-feeding insects arise as a result of reduced hunting pressure. A diversity of plants growing in the garden at all times allows a maximum chance of survival for insects that help maintain balance.

Creating a year-round garden is a struggle, unless you are an extremely dedicated gardener. But once established, a fully packed garden becomes just about as easy to manage as a single-season garden. The work is spread out more thinly over the year, that's all; and the weeding is actually decreased because all the empty space is taken over by plants lusting to expand in all directions. Whether by sneaky underground runners or by flinging seeds in all directions, the plants you put in will fill in all the spaces. Intelligent is the gardener who lets the garden have its way—just short of anarchy.

The easiest part of the gardening year, at least for many of us, is winter. There are books on winter gardening, and I politely chose to ignore them altogether. There is little I can or want to do in my freezing garden between November and March. All the same, a good garden should provide shelter for all overwintering insects. A good basic design might interweave the entire garden with dwarf conifers and evergreen shrubs such as hollies, rhododendrons, andromeda, false box, real box, nandina, and lavender cotton. These plants provide numerous protected places such as twigs and dead leaves on the ground to serve as overwintering sites for all those insects we so desperately want to keep in our gardens. And we especially need to have the predators and parasitoids ready to spring forth to feast on all the plant-feeding insects. The worst possible scenario we gardeners create with our bare-earth policy is that of decoupling the predators and parasitoids from their prey. We don't want to race into gardening season having to battle the tides of imbalance, there is enough to do already.

In the spring, of course, most gardeners will want to tidy up their gardens. I have no problem with that. Either coarsely cut up the top debris and drop it as mulch or take it to the compost pile. The insects will find their ways in and out of such predicaments. (Those of you who bag up your garden debris every winter or spring and have it hauled away with the trash should also be hauled away and placed in a giant apartment building in the middle of a big city. It's really the best place for people who do not understand how nature works.)

The temporal planning for a garden is basically not as difficult as you might imagine, even for a beginner. One simple and entertaining trick is to peruse a good garden center once a week (or every other week for the lazy gardener) during an entire growing cycle (this will differ by geographic region). Walk slowly up and down the rows of plants and look at what is blooming, but more importantly, look at what is attracting insects. These are the plants you want to buy for your garden. It's a one-stop ecological bonanza. In a year, you will have purchased the most likely plants to attract insects over the period of your growing cycle. With a little thought, these known attractant plants can be worked into any garden design, even one that already has plants.

Another approach to the temporal factor of gardening is to be observant wherever and whenever the opportunity arises. When walking your dog in the park, for example, or hiking along a favorite trail, notice what insects are visiting what flowers. When visiting other people's gardens or while touring botanical gardens, observe and take notes on the plants that attract the most insects. Eventually, you can seek out these plants from local nurseries or mail-order firms. Again, you will be assured that these plants will attract insects.

In terms of temporal structure and the enticement of insects into the garden over a consistently long period of time, I am not advocating the flower gimmick solely for immediate gratification. Attracting insects is what flowers do best—it is the job they evolved to do over millions of years. To put it bluntly, flowers need insects to provide pollination. We gardeners should emulate flowers—not to be pollinated, of course—but to lure insects into visiting our gardens, first to find sustenance, then to find enough shelter to be induced to stay for a while. If some insects stay, then others will follow. We will be encouraging a permanent, low-level diversity that begins to build upon itself. The goal is to build layer upon layer of simple plant diversity until the insect-plant and insect-insect interactions become so complex that they take care of themselves, and we poor, simple-minded gardeners won't need to worry about such things.

The one great thorn in this approach to gardening is that these flowering plants may have great attractant powers for insects, which is good in general, but they may not be the best plants for keeping resident populations of insects in the garden on a prolonged or permanent basis. For this to happen, a garden must have plants that are attractive at flowering time and offer components for other parts of an insect's life cycle. Usually this is a larval food item such as the leaves, stems, or roots of specific plants. If a garden has a diversity of plants, there will be choices available to insects and they may find something to feed on and complete their life cycle. Some insects, however, are restricted in what they eat during one life stage. If you want to entice specific insects, such as butterflies, you must be more scientific about the process and delve into life histories. I will cover this aspect a bit more in chapter 11. But for now, we are looking at general ways to establish increasing diversity of all kinds in the garden.

Garden (Habitat) Structure

As we have seen, there are many ways to increase plant structure, and thus insect diversity, by using a variety of plants over time. Even on a flat meadow we can find hundreds of plant species growing side by side, each finding a slight difference in a particular growing condition that fulfills its needs. These plants vary in size, shape, and flower from the first hints of spring to the first ravages of winter. Imagine what can be done if we also alter the physical structure of our gardens to accommodate even more kinds of plants and insects. The end products might begin to resemble species-rich, environmentally diverse habitats—maybe even habitats like the spaces our subdivisions supplanted when they were built.

Water

The fastest, easiest way to introduce an entirely new world of plant diversity into a garden is to add water. It doesn't have to be the waterworks of Versailles, it can be as simple as a cache pot with a single water-loving plant in it. Anyone can find a place for a pot and some water plants, and it becomes quite addictive if you are not careful. Onto my balcony every summer I haul out five or six different-sized pots of dwarf cattails, Japanese iris, variegated sweet flag, sedges, and chameleon plant (the only way, really, to grow this totally noxious weed). A birdbath surrounded by these plants acts like a neighborhood avian swimming hole. The birds become so absorbed in their ablutions, surrounded by protective greenery, that they do not even notice the cat inside drooling against the sliding glass doors. You can use ornamental cache pots, pans, half-barrels, or anything else that will hold water.

I built my first pond using a rubber liner. I started filling it up with water from the hose and, before it was half full, a few water striders (true bugs) were dancing on its surface. How these insects found this bit of water I have no idea, but within the summer I had damselflies and dragonflies landing on potted cattails emerging from the pond's depths. Within a year the water had a resident population of breeding dragonflies, water striders, and toads. Soon came the frogs, green herons (feeding on goldfish I'd placed in the pond to control mosquitoes), and many other birds who stopped by simply to take a bath in the shallow end of the pond. Nothing brings in the wildlife like water. Interestingly, with the exception of mosquitoes, almost all the insect wildlife is predatory on other insects—especially mosquitoes.

The subject of water gardens is a hot topic these days. There are numerous books about how to make these gardens and for good reason, although few people build one for its insect-attraction powers. Water gardens are built primarily for aesthetic reasons, the power to increase our visual pleasure with more kinds of plant textures and colors. We seldom realize that they also increase the diversity and complexity of our gardens and the neighborhoods we live in, whether rural, suburban, or urban. A touch of water is simply a touch more of nature.

Although a big pond will support more different life-forms than a

Many gardeners covet good drainage, but adapting to the soil in your garden or even altering small areas provides room for a diversity of plants, such as the bog-loving sundew. With sundews, you need not fear that insects will attack the plants, because the plants attack insects—well not attack, so much, as entice them to die on the sticky leaves. Eventually this fly will become one with the insect-eating sundew.

half-barrel, for instance, almost any amount of water is welcome in the garden. Even a wet swale is a grand place to increase plant diversity. No pond is needed to grow cardinal flowers, for example, just a moist place with filtered sun. Monkey flowers need only sunny mud. Gardeners having temperate, shady, wet spots big enough for horses may grow eight-foot-tall gunnera. Joe Pye weed grows in moist spots in the freezer-zone. Do not fear moisture in the garden, relish it. There is so much that can be done in the mud.

Soil

Alternatively, there is much that can be done in sand, gravel, acidic soil, alkaline soil, rocky soil, or clay soil. A garden containing pockets of differing soils produces a diversity of plants best adapted to those soils.

Typically, the gardener's reaction to a clay soil, for example, is to want to lighten it with humus or sand or gypsum. If we have a sandy soil, then we add humus or peat. Often we attempt to standardize or work our soils so that the entire garden area is uniform in structure (tilth) and appearance. Why not accept the garden's soil, or at least patches of it, for what it is? Rather than homogenize the garden to our wishes, we should use the plants that best fit the needs of the soil. This automatically increases plant diversity because different plants do better in different soil types. One great advantage in using native plants is that they fit all the major soil types likely to be found in a garden.

I am fortunate in that my garden consists of patches of greasy clay, a little flood-plane loam, some sandy bits, and woodsy humus (all in less than one-third acre). If a gardener has only one type of soil, for example, loam or sticky clay, then some simple actions might be taken to increase soil diversity by amending a few areas. Humus or sand mixed into a few beds would alter them enough to increase plant diversity. A layer of sand on top of clay or loam (simply laid down, not mixed) will add a new dimension to any garden. In it might grow any number of plants that require quick drainage. On mounds of pure clay, I have added a layer of sand and gravel to create raised mounds upon which to grow alpine or rock garden plants that require dry necks (crowns), but some moisture at the roots. In my wet summers, thymes, for example, grow exuberantly on a bed of gravel, which drains away instantly.

Whether we have gardens of many soil types or but a single type, we gardeners can find ways to adapt the soil for the purposes of increasing plant diversity. In the former instance, we can retain this

Having sandy areas in a garden invites some pollinating bees and predatory wasps to nest and become full-time residents. Different insects prefer various kinds and depths of soils in which to nest. Here a bumble bee, laden with pollen, makes a beeline for her nest in a sandy bed. Her nest is 1.8 meters (6 feet) underground, occupying an abandoned, leaf-lined rodent nest.

diversity to our advantage instead of trying to improve it as we typically want to do; in the latter, we can add pockets of alternate soil types without environmentally demeaning the natural state of the garden. A little variety, in moderation, is not bad.

Exposure

Another element in increasing plant diversity is exposure, that is, how much sun or shade a plant receives. Often this is a factor of the plants themselves. Some grow tall and cast shade, some grow squat and do not. It is common for a sunny garden to become eventually a shade garden by virtue of the shrubs and trees that you plant, never realizing that they will always grow faster than expected. Just as often, this conversion from sun to shade is due to your neighbor's trees over which you have no control whatsoever (excluding the clandestine sorts). Occasionally, a shade garden suddenly becomes a sunny one, as, for instance, when a tree falls over or is blown away. Gardeners are constantly engaged in the battle over exposure, frequently having either too much or too little.

Because there are so many different plants with so many different requirements for exposure, it should be deduced by astute gardeners that the greater the variety of exposure, the more different types of plants their gardens will grow. This varies in part by geographic region, of course, so that primroses planted in the sun in Portland, Oregon, will grow just fine, but grown in Phoenix, Arizona, they would be fried in twenty seconds flat. Exposure must be interpreted as it relates to one's area of residence. But region is only part of the equation, because we can artificially modify exposure, within limits.

Such inanimate factors as fences, houses, sheds, telephone polls, and water towers will have a decided effect on sunlight that reaches a garden and thus the plants you can or cannot grow. Also, the topography of your garden (see Relief, next) defines the amount of exposure you have, and topography can be modified to some extent. A short bush at the top of an earth mound, for instance, can produce the relative equivalent shade of a tall shrub at the base of a mound (providing the mound is a big one, of course).

Because exposure is constantly changing from sunup to sundown, from day to day, from month to month, it is not entirely simple for the gardener to manipulate this factor in any reliable way. Much like attempting to manipulate the insects in your garden, an excess of plant variety may overcome the ability of any gardener to comprehend, let alone plan, the complexities required to diversify exposure. The most fortunate gardener, in my opinion, is the one who starts with full sun and can plan out where the shade might best be placed by careful use of shrubs and trees, walls and mounds, poles and pergolas. Living in a forest requires great fortitude with respect to planning exposure. Felling trees to add a little sun into the garden, thus increasing the variety of plants you can grow, which also increases the number of insects you attract, is most likely to be about the last procedure most gardeners will undertake in the name of biological diversity. With some personal reservation, I understand the reluctance to remove trees. That is why I will be seeking out a veritable sunny field for my next garden.

Relief
Relief in the garden is not an opportunity to sit down and rest a spell. Relief refers to the topography (or three-dimensionality) of a garden, and it may range in nature from flat (a meadow) to vertical (a cliff). You might think that a garden's relief is innately unchangeable, which is right in part, but not always. True, if there is a south-facing, hundred-foot-high cliff in your backyard you basically have to live with it. Still, most gardens, even if level as a billiard table, can achieve some change in topography with a little effort. And any alteration in ground level will increase the diversity of plants that can be grown.

The easiest such change to imagine is a simple mound. Soil mounding (and sculpting or contouring) offers a basic opportunity to add different soil types (if desired), orient plants to different exposures, and create islands of diversity in what otherwise might be a sea of uniformity. When mounding, the raised area can be oriented in an east-west direction, for example, thus maximizing a southern exposure for plants that need lots of sun and heat, while producing a protected northern bank for plants that prefer shade. Thus, on a flat plane both extremes can be accentuated and the diversity of plants enhanced even when using the same soil type as found at the site. If a different soil type is used, plant diversity can be enhanced even more. Mounding also offers the opportunity to use plants that require good drainage provided by the elevated soil, so you can create mounds next to bogs or ponds, thus growing plants that require no drainage and those that require excellent drainage within short distances of each other. The creative possibilities of mounding and contouring are endless, and the resulting changes in soil, plant position, and exposure create an opportunity for vast amounts of plant diversity. Again, any increase in plant diversity allows a potential increase in insect diversity.

Mounding is a simple technique to add variation in the garden. Similar results could be achieved by various degrees of terracing or building walls, rock gardens, and waterfalls. Perhaps the ultimate in "mound-think" is to first create a negative mound—a hole in the ground, or what some would call a pond—and use the excavated soil as a backdrop mound behind the pond. This maneuver creates a great deal of relief with a single stroke of the spade and increase the diversity of any garden by about 150 percent.

Mulch and Organic Debris

A lot of emphasis is placed on mulching, especially as a form of moisture retention in warm, dry times and winter protection in cold times. Little is said about the positive role that mulches play in providing a home for insects. Nor is much credit given to organic debris and thatch as refuges for insects. Instead, we are ordered to rake up and destroy all unsightly materials such as dead and fallen stems and leaves, spent flowers, grass clippings, and the ordinary bits and pieces of a garden's life. All the garden books tell us to do this so that "bad things" won't have a place to hide or to overwinter and reinfest your garden in the spring. Well, as I have explained, if bad things don't have a place to hide, neither do good things because they are often hiding inside the bad things. You cannot have it both ways. It is best to keep everything that the garden produces in the garden for as long as possible. It's the greedy gardeners who survive the test of time.

I am going to say something that will make all the messy gardeners just as happy as pigs in a mud puddle and all the garden writers mad as hell. Don't be so tidy. It is not necessary to cut down and cart away all the so-called yard debris that litters our landscape. If you must cut stuff down, rough chop it and tuck it at the bases of plants. If the ground is covered by plants (as it should be to keep away weeds), then simply hide the debris under the floral rug. If the ground is open and barren, use chopped up roughage as mulch (it will soon disappear) or mix it with pretty mulch that you buy at the store. Many herbaceous perennials and grasses can keep their dead tops for the winter. Leave some seedheads of herbaceous plants for the birds instead of buying high-priced seed at the supermarket. If none of this sounds appealing, then rough chop the dead plants and put them in the compost for use next year.

Current studies show that many predatory insects make their homes in this sort of twiggy, thatchy, mulchy environment. These areas act as refuges for prey and predator alike, as well as reduce cannibalism and direct competition between predators who might feed on each other rather than on plant-feeding insects. This environment is particularly inviting to predatory beetles and spiders that move up and down the associated growing plants looking for hosts. In general, adding (or retaining) mulch, organic debris, and thatch is good for the garden and helps to stabilize species populations. It is also an efficient use of organic material, not to mention a good excuse not to clean up the garden too thoroughly.

Building a Garden of Diversity

A fairly safe piece of advice that might be given in the context of increasing garden diversity is that it is important to try to view the garden as a living structure that exists within the natural world, not apart from it. The dilemma that becomes immediately evident is

Fallen fruits, berries, vegetables, sticks, leaves, tree trunks, and dead plants all become food for passing insects, bacteria, and fungi, which conspire to break down the organic matter to its principle elements. Here a red admiral butterfly sucks moisture from a rotting apple.

that the natural world is not all that apparent anymore—just ask anyone living in the middle of a city or even a suburb, for that matter. If we gardeners are to have diversity in the garden, it must come from somewhere. This point illustrates the importance of all those native-species folks who are popping up like weeds in all manner of spaces. They have at least one extremely serious point that should be taken to heart by all of us herbaceous-border infidels.

Earlier I suggested that we could build an artificial community of plants that might mimic a natural plant community. Such a community—let's call it a garden—might mimic a natural community, at least in part, by having an ecological complexity and thus an inherent balance. I chose my words carefully and did not say we can build a "natural community of plants," because I do not believe that gardeners can do this. We cannot build a natural community in the sense of it being composed entirely of native (or naturally occurring, local) plants and their insects. The reason for this is that gardening, by its very nature, is the antithesis of being natural: gardening is managed care, whereas natural is not only benign neglect, but downright lack of interference.

If gardeners are serious and interested in returning to some sort of earthly (or even earthy) paradise, we can attempt to create an ecological homologue, a reasonable copy of our immediate natural surroundings (be it desert, rain forest, or alpine meadow). This would be composed of plants that look like those in the surrounding plant community, but may have come from anywhere in the world. A good example of this is what might be found in a Mediterranean-style garden (one with hot, dry summers and moderately cool, wet winters). The reason that many temperate parts of California, for example, look like parts of Australia, South Africa, or Spain is because plants from those regions grow perfectly well in any area that has a Mediterranean climate.

The basic problem with this sort of homologous garden is that, although it might attract many insects due to the similar bloom times, floral structures, and so on, it likely will not induce insects to stay because they will not eat plant material that is totally foreign to them (which is often caused by unacceptable chemical compounds found in these plants). The most reasonable approach that gardeners can take is to integrate some native plants into our garden schemes. This would ensure that when insects are drawn into the garden (from local, relatively natural areas), some of them, at least, will find a preferred food and will be induced to stay. We must decide at some point how generous we are going to be to our fellow animals on the planet. Are we going to invite insects into the garden for a quick visit—you can sip the nectar but you cannot stay? Or are we going to insist they move right in and take up their own tiny battles of survival, balance, and stability?

The decision on how and why each of us gardens is a matter of personal taste, philosophy, emotion, artistic disposition, natural stewardship, maternal instinct, environmental concern—the list grows ever onward as each of us finds our own purpose for being in the garden. We gardeners are the people who most desire to bring plants into our lives, but by so doing we alter the very balance that existed for eons in the physical spaces our gardens now occupy. Is it not reasonable that we attempt to embrace all life-forms in our gardens in an effort to restore the balance that the garden, itself, has displaced?

11

Inviting Insects into the Garden

There is currently a great movement afoot to invite one special group of insects into the garden. Naturally, it is the group with the most beautiful, most innocuous members—the butterflies. Mind you, we still do not invite the butterflies' second cousins, the moths, into our gardens, but at least we are starting to accept the butterflies and even actively encourage them in many places. After butterflies, however, the number of insect guests drops precipitously to nearly nothing. I will now give some idea of what is going on in the world of formal, engraved invitations to our friends the insects, but it will not take long to do so.

Insects as Guests

When it comes to garden visitors, there are two basic principles involved in the art of the invitation: one is the simple drop-in-for-a-visit arrangement and the other is the set-yourself-down-and-stay-a-while variety. In the first case, we are offering appetizers or drinks for the drop-in insects—adults mostly. We are willing to liquor them up with nectar, but then we hope they will go to some imaginary home before causing any possible damage to our garden. A butterfly, for instance, comes to guzzle at a flower, but it does not stay much beyond the sipping phase because it does not find any suitable plant food for its young. Butterfly adults consume mostly nectar and to a minor extent pollen, but butterfly larvae eat plant parts, the sort of things we gardeners don't like to see eaten. Whereas adult butterflies may drop in for a sociable visit, their children will want to stay a few weeks and bust up the place if there is anything good to eat.

The gardener must decide what it is that she wants for her gardens. If the invitation is for guests to drop in and stay on a permanent basis, thus paying for their room and board either by their elegant good looks (butterflies) or their practical virtues (predators and parasites), then these insects will need the resources that provide staying power. This includes host plants for butterfly larvae to eat and plant-feeding insects (what everyone calls pests) for the predators and parasites to eat. The gardener will have to tolerate some potential plant damage in return for a degree of guaranteed visitor permanence. What this means, in practical terms, is restraint from reliance on chemical defenses.

If, however, all the gardener desires is the presence of pretty little things such as butterflies or what we tend to think of as "good bugs" such as ladybugs, then he may be able to attract them by cunning (for example, nectariferous flowers), but this cunning will have essentially no lasting effect. By virtue of their natural inclinations, insects will need to find sustenance for their offspring; they may pass our way, but they will not stay in our gardens if there is nothing for their young to feed on.

There is one additional consideration in all of this general inviting, however, and it has to do with the degree of naturalness surrounding the garden in the first place. If the gardener lives in a patch of cityscape, for instance, it is likely that the number of different insects passing by is already greatly diminished. Because natural areas are not typically retained in cities, there are relatively few reservoirs of plants in which insects can breed and conduct their lives in what might be considered a natural manner. The insects will be mostly tramp and cosmopolitan species, those that live in easy association with humans. So here sits the urban gardener—the

Many gardeners would be satisfied simply to have one of the great black-and-yellow swallowtails visit the garden on a regular basis.

butterfly gardener, let's say—with an open banquet table and no guests, or at best a few stragglers that wander by looking for the promised land.

I make these points not to discourage, but to say that the ideal of insect gardening is a bit more complicated than simply planting flowers that attract insects, whether butterflies or other insect guests. These insects have to come from somewhere, at least the first time, and the garden may not be situated near any land natural enough to provide only the insects one wants. In fact, in suburban and urban gardens, we mostly end up with the insects we don't want because they are the weeds—the hardiest of the hardy—of the animal world.

Butterflies

There are more than two dozen books, by latest count, on the apparently specialized subject of butterfly gardening. I will be honest and admit that I have not read a single one all the way through. It goes against what few principles remain in my life to encourage only one group of insects in the garden. Because most butterflies are limited in distribution by the plants on which their larvae feed, butterfly gardening is essentially a regional topic. It is difficult, therefore, to make all-encompassing statements that answer questions associated with specific regions, doubly so when it comes to the species of butterflies likely to be attracted and what their host plants might be. In this section, I give an overview of the basic philosophy of attracting butterflies, but leave you to do the research necessary to determine the specific plants you might need for your area.

To attract butterfly adults, you need plants. Experts tell us that nectar-producing flowers in shades of purple, yellow, orange, and red are best. If these flowers are tubular, so much the better. Butterflies have long tongues that uncoil to reach into long, tubular flowers, and it is easier for butterflies to extract nectar from single flowers, rather than double flowers. Also, large patches of a single color are easier to spot than single flowers. Butterfly gardens should have overlapping bloom so that the garden offers a long period of attraction to many different kinds of butterflies.

The garden needs several structural elements. There should be sunny and shady areas and a diversity of plants to produce a variety of microhabitats and bloom times, which increases the gardens seasonal diversity. The adults require areas such as flat rocks and open spaces to rest on and wet, muddy spots for puddling. Butterflies love to suck moisture from mud—perhaps they are the pigs of the insect world. Any self-respecting butterfly collector will seek out mud puddles (especially in the tropics), at which huge aggregations of

Monarch butterflies are sometimes common in the late summer as they begin their long migration back to Mexico from points north.

many species of butterflies may be found. These aggregations may number in the hundreds or thousands. What's good for the collector is good for the gardener.

When butterflies have found a sufficient food source, they mate and begin searching for host plants on which their larvae can feed. Whereas adult butterflies will feed on almost any accessible nectar-producing flower, their larvae will only feed on a few select plants. These plants are determined by the butterfly species, and adult female butterflies will seek relatively specialized hosts for their offspring. Therefore, to gather the full benefit of attracting butterflies,

Skippers are one of the most common butterflies in our gardens. Perhaps that is the reason we take them for granted. Skipper larvae are grass feeders.

larval host plants should be included in the garden. These would be plants native to your area or at least near relatives of these plants.

Without going into great detail on the subject, here are a few of the more common butterflies and the host plants upon which their larvae feed: Baltimore checkerspots on turtlehead; black swallowtails on apiaceous plants, such as dill, fennel, or parsley, and on rue; eastern tiger swallowtails on tulip tree, wild cherry, and lilac; field (sachem) skippers on Bermuda grass and other grasses; gulf fritillaries on passionflower; monarch butterflies on milkweed species; mourning cloaks on willow, aspen, and elm; pipevine swallowtails on pipevine; silver-spotted skippers on locust and wisteria; spicebush swallowtails on spicebush and sassafras; viceroys on willow and cottonwood; and western tiger swallowtails on wild cherry, sycamore, and willow.

In butterfly gardening, as in all of life, one must attempt to be realistic. A gardener does not build a butterfly garden in one day or even many years. In, fact, according to Jeff Cox (1991), research done in England indicates that even a one-acre backyard wildlife sanctuary scarcely housed the life stages of more than a few different kinds of butterflies. In a one-fourth-acre suburban backyard, three species of small butterflies were found to breed, but more encouraging was that over a five-year period 9000 butterflies, representing thirteen species, were captured, marked, and released from the same yard. Obviously, with dedication and some knowledge of local butterflies and their feeding habits, a lot of activity can be generated in a small space.

One final note on butterfly gardening, and I then I will proceed to the really interesting sorts of insects you might want in your garden. As with all fads, there is a tendency for the commercially greedy to prey on our hopes of enticing butterflies into the garden. The notion of building—or even worse, buying—butterfly houses (also referred to as hibernation boxes) and placing them in the garden is founded upon the caring attitudes of the gardening set. Pay it no mind. Hibernation boxes are utter nonsense, in fact, they seem

The blues are small, intensely beautiful butterflies. The larvae of some species of blues live in ant nests, where they are protected and fed in exchange for the honeydew they secrete.

Copper butterflies are related to blues (see previous photograph). They are equally small and equally beautiful. Because they feed on the weed dock, most "good" gardeners will never have coppers take up residence in their gardens. What a pity.

almost to be deadly. In a study undertaken at Penn State University, forty butterfly boxes were placed along woodland trails where over-wintering butterflies were commonly seen. Over a year's period, not a single box held a butterfly. More than half the boxes held spiders, however, so the boxes could better be thought of as butterfly coffins.

Clear-winged moths fly much like hummingbirds and are a virtual blur for most of their lives. Their larvae bore into roots, stems, canes, or trunks of plants and trees. The infamous squash borer is a clear-winged moth. If all we saw were the adults, we might invite them into our gardens along with the butterflies.

Bees and Wasps

With the words *bees* and *wasps* comes the compulsory throwing up of the hands and the obligatory screaming. Let it all out, folks, and get over it. With the exception of people who are deathly allergic to bee or wasp stings, almost no one need fear these insects. In this section, I will tell you how to attract these good bugs. If you cannot seem to come to grips with this notion, then I suggest you turn straightaway to the next chapter and read up on the art of losing your fear. Maybe then you'll be able to return and finish this section.

You may recall that solitary bees are valuable pollinators and that it is our civic duty to protect these little fuzzballs. You may also remember that wasps, especially the ones that sting, are one of the garden's super predators—able to grab a gypsy moth caterpillar in a single leap. It seems only logical that these beneficial groups of insects should receive as much good press as do the more glamorous (and certainly less useful) butterflies. But, of course, they do not. It's negative press or nothing for the sisterhood of stinging wasps, bees, and ants, with one minor exception.

The honey bee has been mollycoddled by humans essentially since the first person accidentally, we presume, stuck his hand into a hive for some reason or other. Naturally this person got the bejeebers stung out of him for doing so, but in an attempt to ease the throbbing pain in his hand, he licked it and discovered the wonderment of honey. Humans have worshipped the honey bee ever since. If we don't steal the bees' labors outright, we build them straw or wooden empires, turn them into tenant farmers, and then steal whatever we want from their hives. Businesses that sell bee equipment and rent out bee colonies, especially to commercial fruit growers, have become the norm. For anyone who wants to raise honey bees in their garden, the information is out there in books, magazines, telephone directories, the Internet, and probably any Eagle Scout you'd care to confront. Because so much has been written about them and because honey bees are completely foreign to North America, I will not discuss their introduction into the garden. You

272

I find it difficult not to like bumble bees. They are furry, funny, and dedicated to one thing—work. It's difficult to make a bumble bee mad. I only know one person who was stung by a bumble bee. A queen decided to make a nest inside a boot that was sitting on the back steps of the house. My friend put her foot in the boot and was stung. It seemed only fair to me.

will likely have honey bees whether you want them or not, and, let's face it, a big, buzzing honey bee hive is not what most folks want in their gardens. It scares the neighbors and any people you might want to visit. No, it is far better to have the solitary bees, rather than those conquering invaders, what some entomologists call "pollen-pigs," in reference to honey bees' totally unnatural, farmyard-like relationship to humans.

The general principle of attracting a solitary bee is akin to attracting an adult butterfly visitor—a quantity of flowering plants of all kinds. But unlike most butterflies, the bee visits the flowers, in part, to gather pollen. The adult bee does not need a food plant on which to lay its egg, as a butterfly does, instead it needs pollen, which it places into cells and on which it lays an egg. When the egg hatches, the bee larva feeds on this stored pollen, pupates, and then emerges from its cell as an adult. What is necessary here, in addition to pollen, is a suitable place for the bee to create a cell, and this place varies from species to species.

Solitary bees have several basic preferences for nesting. A few species build resinous or pebble-covered nests under rocks, but these are not commonly seen. A great many more species burrow into the soil, some of which prefer hard-packed, clay soil approaching the consistency of concrete, whereas others nest in loose sand. Some species prefer open, horizontal areas such as dirt roads or sandy beaches (above high river or tide), whereas others build nests in vertical cliff faces such as those along riverbanks or crevices in rock walls. Some species burrow into vertical wooden surfaces such as dead trees, fence posts, old barn doors, and the like, whereas others burrow into the ends of dead twigs. Many bees, however, are opportunistic and use the abandoned burrows of previous tenants already present in the wooden structures. Where clapboards are overlapped by vertical endboards, bees often find the tiny V-shaped openings a suitable place to nest. In old buildings or those not well cared for, it is common to see bees investigating what appear to be openings, but are really nail heads that have lost their paint covering. (Bees may be industrious, but no one would say they are especially smart—except maybe honey bees, who communicate with each other in the dance language.)

Bees prefer to nest—whatever the substrate—in areas that do not suffer constant agitation. They do not like to be disturbed while excavating a nest, for example, and the larvae cannot survive if the ground they are in is spaded of forked over every few days. Bees would be likely to nest in a pile of sand or mound of soil that some slothful gardener put in the driveway and did not move for a year. I have encountered nesting bees (and wasps) in both situations when I've mislaid piles of sand and soil for a couple of years. One summer I had a colony of small bees (scarcely noticeably to me and likely invisible to the untrained eye) nesting in an outdoor flowerpot containing a root-bound banana plant in hard-packed soil. This pot was on a second-story balcony. The bees did not seem to mind the frequent watering they received and continued to nest all summer in full view of anyone who could see the dozens of tiny creatures going about their business. To them, a pot full of soil raised high off the ground was nearly an entire universe (they had to go elsewhere for pollen, but the principle is the same).

As with the butterfly fad, there are a few enterprising (and decidedly more realistic) entrepreneurs designing and selling products to attract and keep solitary bees. Because many species nest in openings in wood—either milled or still in tree form—it is easy to trick them into doing so with blocks of wood that have been properly drilled. Different species of bees prefer holes of various diameters (correlated to their body size), but typically one block of wood is drilled with all the same sized holes for attracting one species of bee.

Somewhat ironically, solitary bees are a bit gregarious in that when a female emerges as an adult from her own larval cell she will almost certainly search the immediate vicinity for a suitable hole in which to begin a new nest. (This is especially true for soil-nesting bees that excavate their own nests.) After a few years, if left undisturbed, several females may be working in the same vicinity. If the nesting site is good, a nesting aggregation may form. These bees are still solitary in that they do not interfere—or even interact—with each other; they do not cooperate with feeding or with anything, really. Instead of the highly organized office building of the honey bee, these solitary bees may end up packed together like residents of a townhouse—neighbors in building only.

Perhaps of special interest to some, myself included, is housing for bumble bees. Although these bees are absurdly common, I'll bet there is not one person in ten thousand who ever saw a bumble bee nest. I've only seen two natural ones in my entire life, which is now getting on past the half-century mark—well past, some would say.

Tiny bees pollinate tiny flowers. This small carpenter bee, a solitary sort, nests in dead plant stems. Leaving some debris in the garden encourages many such solitary pollinators and predatory wasps to stay and work for free.

Oddly enough, one of these nests was discovered in my garden only this year, the very same year I purchased an expensive wooden nesting box to lure these bearlike bees into my garden.

Bumble bees are social insects that live in colonies in the same manner as honey bees, but their nests are entirely different in internal structure. Bumble bees prefer to nest in the ground in abandoned mouse burrows. I am told by those more knowledgeable than I that bumble bees have an affinity for the smell of a mouse and they are attracted by the comfy nesting materials soaked in mouse odor. My colony, however, had not heard of this mousy predilection. It decided to spite me and ignore the perfectly good—and perfectly expensive—housing I had provided, choosing instead a birdhouse that had been hanging in my garden for some years. These bees, perhaps sensing that I was a professional entomologist and desiring nothing more than to embarrass my book-learned knowledge, placed their nest eye-height off the ground in a place bumble bees do not normally nest. This is a perfectly good example of why it is best to leave all the thinking to insects and simply provide them a smorgasbord of opportunities to invade the garden. No matter how or what you do for them—no matter what sorts of flowers you plant, how many ladybugs you release, or how many nesting boxes you place out—these insects will do what they damn well please. The best the gardener can do is to offer everything under the sun and hope for success.

I have said nothing yet about attracting wasps, but the principle is virtually identical to that for bees. Wasps come in the social varieties (mostly yellow jackets, paper wasps, and hornets) and in endless solitary varieties. With few exceptions, all wasps are hunters. The chances are high that no one will want to encourage the social wasps, and I basically agree with this. There have been attempts to manage yellow jackets in the same manner as honey bees, primarily for agricultural purposes—several colonies of hunters flinging terror throughout the cornstalks is a practical idea for some, but not for the gardener. Without our help, these social wasps find their way into our gardens just fine, picking off caterpillars here and there and often nesting within sight the entire summer without our knowledge. We think nothing of them, as long as we don't see the nest. But as soon as we do, some killing instinct in our psyche arises and we douse the whole colony with gasoline or noxious chemicals. I will not belabor the point, but even I would not insist that you make houses for the likes of yellow jackets or hornets. Alternatively, I simply ask that when you see a nest already in place, consider the total biological picture before you run for the gas can. In mid to late summer, if you find a nest next to the pathway, chances are it was there the whole time and didn't bother you at all. You shouldn't think just because you have found a nest that its occupants, by some strict code of wasp ethics, are now required to kill you.

On the subject of solitary wasps, however, I do insist that you allow these maligned creatures a chance to live in your garden. Different species nest in precisely the same way as bees, so whatever one does (or doesn't do) for bees will be just fine for wasps. Solitary wasps will even use the drilled-block houses or sticks that are made for bees.

Being predators, wasps are fairly well matched in size to their prey. Very tiny wasps, for example, will attack and kill little insects like thrips and aphids. Bigger wasps will hunt spiders and caterpillars; even bigger wasps will bring down cicadas. Most of these wasps nest in the ground, in twigs, or holes in wood and perform an important function in the garden (insect control) with scarcely a notice. Chances are the average gardener will not have seen 99 percent of the hunting wasps living in the garden, and that is probably a good thing. We might otherwise kill them outright and not ask questions.

Other Insects

Having discussed the butterflies, bees, and wasps, to my knowledge, no other insect has been purposefully invited into the garden. No books have been written, for example, about beetle gardens or dragonfly gardens. Certainly no one has penned a treatise on earwig or grasshopper gardening. There are no books on building thrips houses or fruit fly apartments. This is a pity because, as we now know, all insects should be invited into the garden, if only to take care of each

Few people go out of their way to invite such beautiful creatures as this dragonfly into their gardens. All a gardener needs to do is add water, and the dragonflies will come.

other in their ordinary predator-prey interactions—the natural way, that is.

A few miscellaneous things have been written about how to attract insects and how to construct houses for them to live in, but these instructions generally come with the idea of trapping insects with the secondary implication of butchering them. Yes, let's just be honest and say that we invite many insects into the garden simply to do them in.

The infamous rolled newspaper is one such invitation designed to entice earwigs away from our plants, but this is more an earwig slaughterhouse than what I would call an invitation to mutual co-existence. The Japanese beetle trap is another such device. And the electric bug killer is decidedly brutish. Sure, extend an invitation to the poor, homeless blighters and then make certain they never leave. Really, such lack of heart and neighborly thoughtlessness is beneath a gardener's dignity.

As I noted earlier, it is just as likely that earwigs are beneficial (by eating slug eggs) as they are harmful (by presumably eating plant parts). Attracting Japanese beetles to your own garden may actually mean you end up with more beetle problems than you would have had if you let them fly about at random, hitting all gardens evenly. As some waggish entomologists have said, if you want to control Japanese beetles in your own garden, persuade your neighbors to put the traps in theirs. Then the beetles fly away from you to them. There is something to be said for this advice. The electric bug killer, sold as a magic cure for mosquitoes, actually kills all night-flying insects except the pesky little drillers. Studies have shown that attraction to light and subsequent electrocution is not an attraction for mosquitoes, but it does bring in thousands of other nocturnal insects, especially moths. Do we really want devices that kill everything except what we think we want to kill?

Posting an engraved invitation for insects to live in our gardens is about as environmentally friendly as we can get these days. It is akin to inviting native plants into the garden (which will have their share

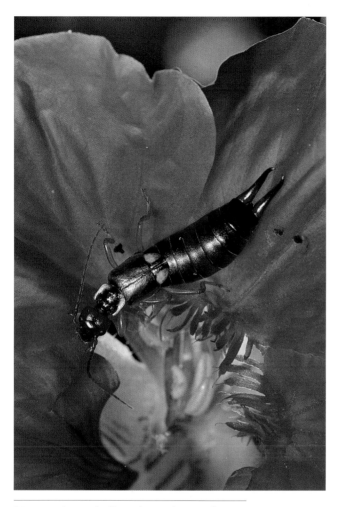

Most gardeners believe that only a madman
would invite earwigs into his garden. Perhaps
we would tolerate earwigs if we thought
they were predators, feeding on the eggs of
other insects. Well, it turns out they are, at
least in part. There is a lot we don't know
about insects and how they feed. Perhaps if
we simply invited them all, the insects could
figure it out for themselves.

of native insects feeding on them, by the way). There are many good reasons to invite the insects, all of which are related to the interconnectedness of the natural world, its balance. Whether the gardener invites insects in by careful planning or by pure, blind luck, the object is to provide an opportunity for a passing insect to stop in for a moment or, better yet, for a generation or two.

12

Fear and Loathing: A Gardener's Guide

The fear of insects, especially of being stung (or bitten), is one of those quirks of human behavior for which we offer little rational explanation. We simply don't like those creepy, crawly things. Let reason be damned—it's part of the-only-bug-is-a-dead-bug mind-set. Although there is a clinical psychological disorder called entomophobia (arachnophobia, in the case of spiders), in which a person suffers from an earnestly irrational fear of insects, this certainly is not what troubles most gardeners. As far as I can tell, the things that trouble most gardeners about insects fall into two categories: first, that a bug is going to do some unimaginable harm to the gardener; second, that a bug is going to do inestimable harm to the garden. In this section, I try to present a realistic view of the natural world and our artificial world, the garden.

Gardeners in Crisis

A few years back, I was on my knees edging the lawn. The process was the standard technique of pulling the longer grass with my left hand and snipping the shorter grass with my lawn clippers. Suddenly, my left hand felt a sharp stab of pain, as if I had been slashed. The pain was so acute I thought I had surely run my palm over a discarded razor blade or perhaps a piece of broken glass. When I pulled my hand up, there was no great stream of blood, as I expected, but my palm throbbed as if someone had hit it with a hatchet. Although I wanted to know what happened, the pain was intense and would not stop, so I went into the house and bound my hand in a towel full of ice. This helped, but not enough, I must admit. By then, my mind was so preoccupied with whatever hideous demon lay in the garden that I ignored the pain and went back to view the spot where I was working. I poked about, timidly I should add, where I was injured and within minutes found the culprit. It was a moth larva, a caterpillar of the saddleback moth to be precise.

This larva is quite attractive, with lime green ends interrupted in the middle by a brown saddle outlined in white, but it is also covered with what we entomologists call urticating, stinging, or nettling hairs. These hairs break off and a poison fluid is released into the enemy (or in this case, me). The intensity of pain is worse than that of a plant nettle and much longer lasting, which I know from personal experience, having on occasion mistakenly walked through patches of nettle.

As part of the hazard of working with bees and wasps all my adult life, I have experienced stings ranging from the puniest of bees to the most magnificent of tarantula-killing wasps. I have experienced pain from the soles of my feet to the roof of my mouth, as well as a few very tender spots about midway between the two. I can assure you that saddleback caterpillars are right up there with the best of them, coming in just after the sting of velvet ants and harvester ants, which, at one time, I had considered to be the most excruciating pain I'd ever experienced—the kind that makes a reasonably sane man cry.

Is there a point to my confession of carelessness? Why, yes, there is. I am still alive (although some have argued the point), here in my study, writing about the painful facts of life. That is to say, after being in harm's way nearly my entire adult life as a professional stalker of wasps, never once have I died. As a gardener my whole adult life, the saddleback caterpillar is the first faint brush I've had with a near-death experience in the garden. I think gardeners can take comfort in the fact that barely a single one of us has ever died from insect aggression in the garden.

As a biologist, I am compelled to use the phrase "barely a single one" because the statistics on garden-related insect deaths are not readily forthcoming. I have read that about forty persons a year die in the United States from bee and wasp stings, or more properly from allergic reactions to these stings. But we know little about these statistics, even if they are true. We do not know, for example, if the victims were pulling weeds or snoozing in a hammock (gardeners, of course, should never be caught snoozing when there is work to do, but things happen); if they were trimming the hedge or pouring gasoline on a wasp nest; if they were planting dahlias or walking through a botanical garden. We only know that about forty persons a year are unfortunate enough to meet their untimely end from a bee's end and that these data have never been correlated specifically to gardening events. Now, should we fear insects based on an absence of information?

If I were a person who was allergic to bee or wasp stings, I might think seriously about gardening and its possible negative consequences. But then, isn't one as likely to be stung driving down the street with the car window open as working in the garden? In fact, I would be willing to bet that not a single person who drives ever considered not driving based on the likelihood of being stung.

Is there truly a need to fear insects? Could this fear just be a primitive piece of behavior left over from our knuckle-walking days— times long ago when we robbed honey bees by cutting down their trees or when we unconsciously stumbled into a wasp nest while grubbing for our supper? A primeval fear of insects might be a basic mechanism of aversion toward things that could harm us. But many other animals, including primates, are content to have insects living and walking all over their bodies as if used to it. The most excitable reaction such animals might show is a little chattering or mutual nit-picking among the clan. We humans, however, clearly do not like the thought of an insect walking on us, much less living on us.

We do not like bugs living *with* us, either: not ants in our pantries, not termites in our walls, not moths in our cereals, not fleas in our carpets. Most likely we resent the fact that we cannot see most of these creatures and thus do not really know what is going on around us. We just suspect that we are surrounded by insects. When we do see them, we somehow perceive insects as bad things. No matter that the insect is cleaning up an untidy little corner of the house—removing crumbs, for example. We simply cannot tolerate things living with us, and it becomes only prudent to take an attitude of better a dead bug than one we cannot see.

Clearly, we have every right to dislike bugs that really might hurt us, by a sting, for example, or, more insidiously, by transmitting insect-borne pathogens that give us such diseases as malaria, yellow fever, and plague. But let us not overreact to all insects simply because we do not like or understand the few that intimidate us. Let's be reasonable in our haste to douse our homes, our spouses, and our gardens in fumigants or chemicals that make the minor insect irritations appear like tiny angels by comparison.

Bites and Stings

For the record, bites and stings are not the same thing. Bites are caused by the front end of an insect, its mandibles, but stings are caused by the back end, its stinger. Most gardeners are likely to ignore such pedantics, however, as they dance madly about slapping themselves and spinning senselessly while trying to escape the torment of those invasive insects.

The examples of insects that bite humans are surprisingly few considering the immense number of insects that exist. What probably comes to mind first is the mosquito, followed perhaps by the biting midge (no-see-um), flea, and to a lesser extent the louse. (I do not include ticks in this section, because, as you may recall, ticks are not insects.) These are among the few insects that bite, rather than sting. Again, speaking technically, they don't really bite, they puncture with their sharp mouthparts and suck blood through straw-like devices. These hypodermic needles are simply modified mandibles that no longer can chew.

I will freely admit that mosquitoes and biting midges are a nuisance. I have them in my garden, and at times I curse them rather freely. I rely on fish in the ponds and bats in the air to help solve the problem. Of course, the pond is what attracts these flies in the first place because they need water for their larval stage. If you have a pond, you are simply inviting these flies to move in. This might be a good thing, however, because in a pond full of fish the larvae will invariably be eaten and the net result may be a lower population of biting flies in the neighborhood. The pond will act as attractant and trap. Theoretically, mosquitoes may be controlled by the use of the bacterium *Bacillus thuringiensis* 'Israelensis', which is sold in the

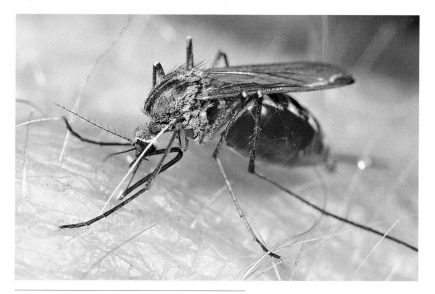

Mosquitoes are among the most annoying insects around, and I am hard pressed to defend them. Perhaps their best defense is that they provide food for other organisms such as bats, birds, fish, frogs, lizards, dragonflies, and damselflies, just to mention a few. Females suck blood through their mouthparts, but males, it should be noted, do no harm at all. They cannot bite, feed only on nectar, and may even provide a service by pollinating plants.

form of a compressed ring that can be flung into standing water. (I have a friend who uses a slingshot to fling these into her neighbor's gutters that don't drain. She claims it works wonders on cutting back the mosquito population. I should have thought to suggest she use her slingshot to fling goldfish into the gutters.)

Fleas live in gardens that have animals, either wild or domestic. There are lots of different kinds of fleas. Some attack squirrels, some attack rats, some attack birds, and some attack dogs and cats. None of these fleas much care for humans, but if dog or cat fleas enter the house along with their companions, they eventually hop off and lay their eggs in bedding and carpeting wherever the animals roam. When the flea needs another blood meal, it hops onto whatever wanders by, and that might just be a gardener. This is not a particularly good thing, but it has happened to the best of us. Because fleas lay their eggs all about and it takes a few days for flea eggs to hatch, the best method of coping with fleas in the house is by frequent vacuuming and whatever flea controls the veterinarian suggests.

By most current standards, lice are rather uncommon insects. They all specialize in feeding on certain animal hosts or even different parts of animal hosts, especially in humans. There are a few sorts of sucking lice that attack humans. The pubic louse is one example and the head and body louse another, which is also commonly referred to as a "cootie," in case you didn't know. The only way that these little suckers are transmitted is by person-to-person contact—no dogs, cats, or gardens are involved. Well, that is to say, no gardens are involved unless one does things in the garden with another person in a way that is likely to involve bodily contact of the sort that is, well, frankly, not the kind of thing one talks about in a book devoted to gardening. Which, thankfully, brings us to the subject of stings.

Some stings hurt, there is no doubt about that. The only order of insects that sting, in the true sense of the word, is the Hymenoptera, which contains the gardener's predatory friends, the wasps; our pollinator friends, the bees; and the detritus collectors, the ants. Such duplicity, both stinging and helping, might incite intellectual discussions about the notion that an insect can be both beneficial and somewhat nasty, but self-preservation would seem to be a natural and easily forgiven attribute.

At least one seemingly good piece of news is that only some female wasps, bees, and ants can sting. That would seem to cut back on our need to fear every living hymenopteran, wouldn't it? All we gardeners need to do is be able to tell a male from a female and we are saved. There is a bit of bad news, though: all the colonies of social wasps, bees, and ants are essentially made up of females. The numerically insignificant males are basically only useful for one im-

The honey bee, one of the most likely of all poisonous animals to be seen by humans, is well known for its stinging ability. It is the only stinging insect that leaves its stinger behind (shown here still attached to its abdomen). This apparatus continues to pump venom into the victim for several minutes after the bee rips itself away and drops dead.

portant function, which is easily guessed, and when that's over their lives are spent.

The vast majority of predatory and parasitic wasps, solitary bees, and colonial ants cannot or do not sting. The legions of parasitic wasps are harmless (with one or two rare exceptions). Even those with relatively long ovipositors (what any right-thinking person would call a "stinger") cannot harm humans; however, they can drill into solid tree trunks in search of embedded host larvae. Most species of garden ants are defenseless (although they may attempt to bite or sting with little effect). Many wasps and bees are either too small or too timid to sting.

The gardener is most likely to find wasp and bee adults paying cordial visits to flowers and plants in search of nectar or prey. In the process of feeding or hunting, the bees and wasps (even stinging ones) are totally lost in the moment and are unaware of almost any other living thing nearby—the gardener included. Even the supposedly vicious sorts of social bees and wasps pay little attention to the gardener who minds her own business. My next-door neighbor, for example, used to keep honey bees in his backyard, and endless streams of them would visit a small pond in my garden to load up on water. For me to stand between the pond and the hives was much like standing on a busy freeway, with the exception that the bees simply skirted around me, instead of running me down. I cannot count the number of times when I've found colonies of underground yellow jackets next to the pathway, the garden tap, or even the back door in late summer. Of course, I was immediately alarmed, but I had been traipsing by these nests all summer with nary a notice taken on the part of either the wasps or me, and so I left the nestlings to live out the remainder of their days.

Where most gardeners get into trouble is when social wasps and some ants (particularly fire ants) are nesting in an odd place, such as a lawn, where they can be easily run over by mowers or children. Colonial wasps, bees, and ants are extremely protective of their houses and their own children (and rightly so). An act of aggression on their house means retaliation. It is a simple notion, really. It is no

A terror of the garden, the yellow jacket is a social wasp (along with hornets and paper wasps). For the most part, gardeners don't even know these predators are around, but disturb the nest badly enough and there is hell to pay. I have wandered past nests in my garden, in complete ignorance, without the least worry. It is only when I become aware of the nest that the wasps seem to become aware of me.

wonder, then, that when someone runs over their territory with a machine or a foot, some agitation arises. This can only blossom into warfare if gasoline is poured into the nest or stones are thrown at it. Many children, myself included, have heaved a bunch of stuff at a nest entrance just to see what the wasps would do—and then been doubly sorry to find out!

Certainly, a wasp nest in an inconvenient place can be a real hazard. Coming upon such a nest gives the gardener pause to ponder his tolerance of all things gardening, including the concepts of good

The cicada killer is a frightfully large wasp (up to 5 centimeters [2 inches]). As with most predatory wasps, females sting but males do not. The wasp shown is a male. Males patrol a territory and inspect everything that wanders into view. Rare is the gardener who can stand still as the male zooms in, at eye level, to get a good view of its trespasser. I've had these wasps living in my garden for several years, and I still duck whenever they fly into view.

and evil, natural and unnatural, right and wrong, life and death. In the simplest view, a wasp nest stands as the extreme symbol of tolerance for all aspects of nature. If it can be said that a wasp nest is in the wrong place in the garden, then a gardener simply has to come to terms with what is the right place. Is there one? In considering this question, the gardener takes a personal position on the character of the garden and even the ultimate disposition of nature.

Gardens in Crisis

The traditional gardener approaches her garden much like a lion tamer in a circus: commands must be obeyed, all things must be in order, must take orders, must follow orders. With this approach, the gardener's role is to demand absolute perfection, a perfect performance every time and every place in the garden. The problem with perfection is that the natural world is full of imperfection—starting at the genetic level—created by countless combinations of variables over which scarcely anyone has any control. In effect, nature is organized chaos—trillions upon trillions of interactions upon which natural forces select those interactions that work, at least for the moment. And each moment's interactions affect the next moment's. Those gardeners who seek absolute perfection (and control) of their gardens are not very realistic or smart, for that matter. I should know, because at times I have been one of those gardeners.

Here's a recent case in point. I was forced to cut down an eastern hemlock tree that formed one end of an enclosing hedge I had been training for ten years. It was an informal hedge to be certain; somewhat naturalistic if you must know, with odd bits sticking out here and there. I didn't want to lose the tree, but it suffered the ravages of the woolly adelgid, an insect that was introduced recently into North America. If this insect is not controlled chemically, it kills the tree it attacks. Dead. No prisoners. No questions. What alternatives can a gardener use to control a nonnative, invading tree killer?

Well, it basically comes down to two seemingly simple options: do nothing and watch the tree die or do something and hope the tree lives. In the case of this insect, control is remarkably easy. It consists of two staggered sprayings, timed to the reproductive cycle of the insect, using either insecticidal soap or horticultural oil. This may be a simple choice for most people, but not for me. Five years ago, when the insect first appeared, I hoped against hope that some native insect might be able to take care of the problem. None did. Then I hoped against hope that spraying it three years ago would work forever. It didn't. Then I hoped in desperation that I hadn't let the infestation go too far. I had. Then I hoped in panic that spraying it too late would help. It didn't. Then I cut the tree down.

Is there a lesson in this story? Yes, there is. Sometimes things

With its elongated head this male scorpionfly looks as if he could painfully stab anything fool enough to come close. Despite its name, the scorpionfly is quite harmless. The males of some species are ferocious looking, sporting a tail end that mimics that of a scorpion. All appearances to the contrary, scorpionflies feed on dead insects.

happen to our gardens that we don't like. We can live with these things (accept the dead tree), we can fight them with hope (reluctant, half-hearted spraying), or we can fight them with knowledge (spraying in sync with the life cycle) and with the least toxic substance that works (soaps and oils). The simple solution, quite literally, is the application of highly toxic chemicals—the easy way out if you do not care what happens after you've killed all the insects on the plant you are spaying. Rachel Carson wrote *Silent Spring* as an

The crane fly appears to be a giant mosquito and, perhaps because of this, it is sometimes feared. These flies are completely harmless, having no ability to bite at all. Their larvae feed primarily on decaying vegetation in moist or wet areas.

Looking as if covered in snow, these dead branches of eastern hemlock are all that remain of a once beautiful hedge plant. The insects, woolly adelgids, a homopterous member of the true bug order, could have been controlled by the proper use of a dormant oil. Sometimes neither nature nor wishful thinking can take care of problems created by man (in this case the accidental introduction of the woolly adelgid from exotic places).

answer to what happens after you've killed all the insects on the plant. It was not a happy picture. More recently, Mark Winston, in his book *Nature Wars*, aptly stated "Our gravest mistake has been in setting a pest agenda that considers pests a problem that must be controlled rather than an integral part of nature that we should manage in effective, environmentally responsible ways."

So just what is a gardener to do when the garden falls victim to a minor or major catastrophe? Well, one thing is to act like the doctor—first, do no harm. In the case of the hemlock, I could have done something to save the tree. But I chose to do the least amount of garden-related harm (from my viewpoint) that I could: starting with doing nothing. In the past, this philosophy has saved me a lot; doing nothing saves time, work, and expenses and, as often as not, has entirely resolved a situation I first considered to be a problem only to discover that it was not.

If doing nothing seems like too much trouble, then the second step a gardener ought to take when insect catastrophe strikes is to gather information. Consult nursery centers, magazines, books, county and university extension agents, fellow gardeners—whatever information it takes to make an enlightened decision. If you are fortunate, by the time you get the information the problem will already be solved. (Those of you who are paying attention will have noticed that this is merely a variant of my first approach, only with more work.)

A third step that can be taken to thwart insect trouble is to assess the damage before doing any harm. To do this, the gardener needs a point of reference, which is why I suggested obtaining information as a second step. For example, if you find a small hole in the edge of a rose leaf, is that really a problem? What about twenty holes? Holey rose leaves might be bad for a rose exhibitor, but not for a gardener with an isolated rose in a shrub border. These holes are made by a little bee, the leafcutter bee, which cuts the leaves to line cells for its young. The bees help pollinate plants in the garden. Thus, the gardener is faced with the doubly complex problem of how much ugliness will be allowed relative to how much service the bee provides. The answers to these sorts of questions are not easily derived. My own inclination is to let insects take care of themselves and to allow a little more leeway than most gardeners might.

As mentioned earlier in this book, economic entomologists have something they call the economic threshold. It is the point of plant damage at which it is economically smarter to spend money to save the crop than it would be to do nothing and lose it. (If the threshold is passed before it is recognized, then the farmer plows the crop under and takes a loss.) In the flower garden, an economic threshold is not too meaningful—we don't eat roses, after all—but it does translate roughly to an aesthetic threshold. How ugly are we willing to let a plant appear before we do something or before it is too late to do something? This is a subjective judgment, to be certain, and one that must be made based on gardeners' needs and tolerance levels. Some gardeners, however, do have the equivalent of an economic threshold, and they tend to have more direct economic concerns than do us flower lovers. These are the vegetable lovers. If a fear of insects in the flower garden is great and a gardener's tolerances low, then the vegetable garden's life is a veritable warehouse of psychological disarray.

I have grown only a handful of vegetables in my life, simply because I do not have room to grow them and I am not willing to invest $500 to produce $6 worth of tomatoes. My personal knowledge of edible plants runs to those necessities of life such as garlic, onions, turnips, Napa cabbage, potatoes, and strawberries. Therefore, my advice to vegetable growers will be limited in scope, but direct in substance. Vegetables are plants. They follow the same set of principles as flowers, which are also plants. Plants is plants, as they say. The only reason that insects like vegetables is because we plant them in big bunches (just like cornfields) and make them easier for the insects to find. When growing vegetables, we plant exactly what an insect finds most delightful—bunches of the same thing together.

Plant-feeding insects detect the presence of potential host plants by several means, not the least of which is olfactory: they smell volatile substances exuded by plants. The larger the number of the same plant growing together, the stronger the concentration of substances secreted into the air for insects to detect. The area becomes a dinner bell to whatever might want to eat it. For this reason, the idea of scattering vegetables throughout the entire garden has been proposed to decentralize one of the primary cues upon which plant-feeding insects orient (see, for example, Rosalind Creasy's *The Complete Book of Edible Landscaping*).

Apparently, humans have known for a long time that, even when planted in a single space, scattering different edible species of plants in localized gardens within a larger landscape provides protection from insects. Evan Eisenberg discusses the concept of what I would call "invisible gardening" in his book *The Ecology of Eden*. Eisenberg tells of the Amazonian gardens of the Kayapo people, who for generations have practiced intensive but biodiverse food gardening. According to Eisenberg, once these gardens are planted they almost take care of themselves. "They have few pest problems, mainly because they are so small and widely scattered—in time and space—that large concentrations of pests can't build up. They can be left alone for months, and at later stages for years." Although the ideas of edible landscapes and placing vegetables throughout the garden are ecologically sound, I admit that it is not particularly efficient, or maybe even satisfying, for the typical produce gardener who wants vegetables in precise rows. Using an ecologically sound approach to food production, in particular, and gardening, in general, is where decisions must be addressed based on a gardener's needs, philosophy, and levels of tolerance.

If a produce gardener insists on inviting insects into the garden, then she will have to deal with these guests. A variety of methods are available, but here again I would proceed by first doing no harm and then getting to know the enemy. Some harmless mechanical methods have proven quite successful, for instance, traps, barriers, and excluders, such as floating row covers. Mechanical picking is not altogether without merit. Adjusting planting times can help avoid particular insects. There are many so-called organic methods, such as the introduction of biological control agents, natural (what some call biorational) insecticides (for example, rotenone and pyrethrum), or insect hormones (juvenile hormone). Ultimately, there is the old standby of synthetic organic chemicals.

If I had chosen, early on, to use chemical methods to control the woolly adelgids that were infesting my hemlocks, my dead tree would still have been alive today. I waited a bit too long to do so. Other, less patient—or perhaps more intelligent—folks might have taken immediate action and called in a pesticide company to coat the tree, the yard, the air, and the neighborhood with highly toxic residues of oganophosphate chemicals. This would undoubtedly have the saved the tree, which might have been a good thing. Or maybe not. It might have killed a few birds, which, if they were starlings, might have been a good thing. Or maybe not. It might have killed off a number of parasites, which, if they were the right sorts, might have held other insects in my garden under control. That would have been a bad thing. Or maybe not. The problem is we gardeners rarely know the full extent of what we are doing when we do it. The fallout could be good, bad, or utterly indifferent. Gardening is a bit chancy that way.

As gardeners, we each have our own set of values, sense of aesthetics, set of goals, sense of right and wrong, and a set of circumstances under which we must face the reality of our gardens and the principles under which the natural world tries to operate. My sense of gardening has tended to change over the years from the all-controlling, obsessed, lion tamer of a gardener to someone more like the ringmaster of a seventeen-ring circus. The death of a tree or a plant is not as painful as it once was because now the garden is so full of growth that death merely represents an opportunity to add a new bit of life here and there. I have never been much on internecine warfare, generally opting to take the easy way out, which, in my opinion, is cramming as much stuff into the garden as will fit and then letting the garden strive on its own—with only a bit of help from me, of course. Nature is, after all, red in tooth and claw. If the garden can stabilize itself on its own terms, who am I to interfere?

13

Appreciating Insects

If you are as fortunate as I am, you may have hundreds of lightning bugs illuminating your trees on a midsummer's eve—a wondrous sight to behold and truly Christmas in July. In late summer, the sounds of crickets and katydids fill the evening's humid air and remind us to make hay while the sun shines, for in a few short weeks this hedonistic merriment will be over for another year. Dragonflies course through our gardens like flying stained-glass sculptures and challenge us to believe that anything so dashing can be real. And butterflies, what could be more evocative than beauty arising from the beastly caterpillar?

Lightning bugs, crickets, katydids, dragonflies, butterflies, these are the dreams of childhood and the memories of times gone by. What would our lives be without them—either the insects or the memories? Certainly the gardener turns petulant when a plant suddenly evaporates into a pile of bug-eaten holes, but, alternatively, what could be more elegant than a dozen black-and-yellow butterflies carousing around a butterfly bush? Are we willing to trade holes in our spicebushes to have spicebush swallowtails? We gardeners are sincerely happy to reminisce about fireflies and butterflies, but we seem none too enamored with the majority of insects we find in our gardens. Perhaps we insist too much on having our zucchini and eating it, too. Can a gardener have it both ways, all pleasure and no pain?

By now you should know the answer to that question. This entire book has been about respecting the role of insects in the garden as well as in the vast scheme of things. Our garden already tolerates many more insects than we would ever suspect and it does so without much complaint, although sometimes a local dispute may erupt between one species of insect and a plant or two.

For eons, humans have interacted with insects whether we wanted to or not. From apocalyptic plagues of locusts described in the Bible to the nineteenth-century Mormon cricket plague in Utah, from flea-induced bubonic plagues of the Middle Ages to mosquito-induced malarial outbreaks of today, we have had to suffer the insects. Yet, not all insects are as bad as they would first appear. Have we not used the honey and wax of the honey bee and the silk of the silkworm without hesitation? We once used the bodies of the cochineal insect to dye our cloth and now we use its derivative, carmine red, as a coloring in beverages, confections, cosmetics, and pharmaceuticals. Shellac made from a resin secreted by scale insects is still used today. Maggot therapy, the use of fly larvae to separate dead, necrotic tissue from living, healthy tissue, which prevents infection, was practiced from ancient times until World War I, before the discovery of antibiotics. Also, indigenous peoples have eaten—and still eat—hundreds of different kinds of insects from all over the world.

Here, I provide the final pieces of an equation concerning gardeners' acceptance of insects in the garden as well as our acceptance of insects as creatures worthy of appreciation in their own right. We have seen how insects fit into the culture of our gardens. Now we will examine how insects fit into the culture of our lives.

If the combination of culture and insects does not seem to be a logical one, I hope to prove otherwise. Learning about cultural entomology simply requires reeducation, for the most part, because we have been culturally exposed to insects without really even thinking about it. Cultural entomology is a surprisingly vast area of study. Even a brief review of the subject led Charles Hogue (1987) to cite more than 200 papers on the subject, and these papers, in turn, refer the reader to many hundreds more.

Language and Literature

Before words as we know them were written on stone or paper, insects played a part in human language. Whether Egyptian hieroglyphs or early Mayan and Chinese pictograms, insects have been portrayed as parables to mirror our own thoughts and the meaning of behavior and existence. Some of the greatest writers and thinkers have used insects as exemplars of our better and our lesser sides. Aesop (b. 620 B.C.), for example, compared our self-righteous and egotistical behavior to that of the lowly fly: "The fly sat upon the axel-tree of the chariot-wheel and said, 'What a dust do I raise!'" When we are industrious and positive, we are said to be "busy as a bee." In fact so much of our life apparently echoes that of the social beehive that humans are forever being compared to what is, really, a somewhat mindless group of organisms: "That which is not good for the beehive cannot be good for the bees" (Marcus Aurelius, 121–80 B.C.) and "Our treasure lies in the beehive of our knowledge. We are perpetually on the way thither, being by nature winged insects and honey gatherers of the mind" (Friedrich Nietzsche, 1844–1900).

Sometimes, as Mark Twain (1835–1910) noted, we are compared to the industrious ant, although perhaps not always as favorably as with bees:

> As a thinker and planner the ant is the equal of any savage
> race of men; as a self-educated specialist in several arts
> she is the superior of any savage race of men; and in one or
> two high mental qualities she is above the reach of any
> man, savage or civilized!

Insects are used metaphorically for both positive and negative comparisons. On the one hand, Virginia Woolf (1882–1941) said of a writer: "He searches out these butterfly shades to the last grain. He is as tough as catgut and as evanescent as a butterfly's bloom." On the other, Lord Tennyson (1809–1892) spoke of the literary critic as "a louse in the locks of literature." It would seem that the type of insect chosen reflects its symbolic significance as well as its perceived station in the universe: butterflies are high-spirited and good; lice are low-down and not.

Percy Bysshe Shelley (1792–1822) wrote of insects to raise the spirits (but not the eyes) heavenward: "I think that the leaf of a tree, the meanest insect on which we trample, are in themselves arguments more conclusive than any which can be adduced that some vast intellect animates Infinity." Henry David Thoreau (1817–1862), however, cast our spirits and eyes directly to the gardener's feet:

A paper wasp scrapes plant fibers with its mandibles, then chews them and makes a paper nest. Could the work of this humble insect have been what inspired Ts'ai Lun to invent paper making, without which humans would have had a history carved in stone?

Is not disease the rule of existence? There is not a lily pad floating on the river but has been riddled by insects. Almost every shrub and tree has its gall, oftentimes esteemed its chief ornament and hardly to be distinguished from the fruit. If misery loves company, misery has company enough. Now, at midsummer, find me a perfect leaf or fruit.

It is likely that almost every human condition—whether practical or preposterous, philosophical or Philistine—has been analogized favorably or unfavorably to insects at one time or other.

Charles Hogue (1987), a student of cultural entomology, noted that he had collected one hundred titles of modern novels in which insects played a major role; the number of short stories was nearly as long. I suspect this number is rather low, if for no other reason than no single person could possibly track the thousands of fictional works that treat insects in some thematic or metaphorical way. (Certainly the arena of science has devoted hundreds of thousands of papers to the subject of insects, but I don't quite think of science writing as literature.)

I do not propose to relate, nor could I if I tried, the whole of insect diversity in literature. Such works would span the ancient to the modern (*Aesop's Fables* to *A Bug's Life* by Justine Korman), the young to the old (*Charlotte's Web* by E. B. White to *Silence of the Lambs* by Thomas Harris), the traditional to the transformational (*Empire of the Ants* by H. G. Wells to *The Metamorphosis* by Franz Kafka), the whimsical to the wanton (*Alice in Wonderland* by Lewis Carroll to *Lolita* by Vladimir Nabokov), the blandly predictable to the erotically allegorical (*Big Bad Bugs* by R. L. Stine to *The Woman in the Dunes* by Kobo Abe). Even the gardener's own Karel Čapek, whose *A Gardener's Year* is possibly the best gardening book ever written, coauthored a play called *On the Life of Insects* (or *Insect Comedy*, depending on the translation).

Over the millennia, insects have inspired a great quantity of literature, and we humans would be poorer without it. (Well, most of it, anyway.) In both language and literature insects have contributed immensely to the richness and wealth of our communicative abilities. And for that reason alone, we should at least tolerate, if not appreciate, their presence in our gardens and lives.

Myths, Folklore, and Religion

It is not too difficult to see how humans' early encounters with insects might generate mythological concepts involving gods and even creation. The Aztecs, for example, believed that Quetzalcoatl, their hero-god, created humankind with his own blood. To nourish his creation, he turned into an ant, entered an ant colony hidden within a mountain, and stole a single grain of maize upon which all humans could survive. Ants are extremely common in the tropics; the ants and the intricate social behaviors of the colonial empires represented by each nest likely did not go unnoticed. Interestingly, Hindu writings teach us that ants are Earth's firstborn and the Earth itself is an anthill.

Early in Egyptian history, the First Dynasty (about 3100 B.C.) was identified with the sign of the hornet, symbolizing a ferocious and fierce opponent. Other cults of the era were identified with locusts. The lowly god Beelzebub (Beelzebul) was lord of the flies (or the insects, depending on the translation). The scarab beetle was a symbol of the sun god, Khepera, and was worshipped as father of the gods, due in part to several numeristic coincidences and because the warrior-like, armor-clad beetles appeared to emerge spontaneously from a filth-encrusted ball of dung (unbeknownst to the Egyptians, of course, an egg had earlier been laid on or in the dung ball).

In Australian mythology, Bushmen believed the praying mantis, Kaggen, was a god of creation. In Chinese mythology, Tschun-Wan was the insect lord of crop pests. In Hopi legend, the world was created by the Spider Grandmother. The Apache of New Mexico believed that fire came from an ancestral campfire ignited by fireflies.

Many of these beliefs appear as likely to be true as the ones I was taught as a child in the foreboding inner sancta of our local places of worship. If there is a significant difference among present-day beliefs and those of yesteryear, one might say that the beliefs developed were derived in large part from observations of the natural world. These days, about the only folks who really pay attention to the natural world, or what's left of it, are the scientists who study its vanishing life and the few people who still care. Everyone else simply depends on it for daily survival. Even the gardeners and farmers, who should know better, pay scant attention to the natural world, this world in which their philosophies of nurturing should be born.

Insect Singers and Music

As a young lad, I was given a couple of cricket cages. (Actually, it turns out they were really katydid cages, but who knew?) I learned only a few months ago that singing insects were a highly developed pastime in China and Japan, where keeping pet crickets reaches back a thousand years or more. Many books have been written about the various types of crickets and katydids and the gorgeous vessels that people keep them in—vessels made of bamboo, clay, porcelain, and even gourds that have been grown in molds to a specific shape and then decorated with elegant, intricate carvings. Expensive coffee-table books bursting with elegant photographs of extraordinarily beautiful cages have been published (for example, *Insect Musicians and Cricket Champions by* Lisa Gail Ryan). People take this seriously, I might add. I have read that "Chinese cricket fanciers allow themselves to be stung [that is, bitten, of course] by mosquitoes. When they are full of blood they feed them to their favorite pet crickets." Now, this is true devotion, not shared even with our pampered dogs and cats.

Insects not only sing for us, but provide some inspiration for writing our own music. Rimsky-Korsakov's "Flight of the Bumblebee" comes immediately to mind, or perhaps Puccini's "Madame Butterfly" or Grieg's "Papillon." Joseph Strauss honored the "Dragonfly" in three-quarter time, whereas Holtz penned "The Beetle's Dance," Cibulka the "Minuet of a Fly," and Liadov the "Dance of the Mosquito." Vaughan-Williams composed the incidental music "The Wasps" for a play by the same name written by Aristophanes, which

A forked-tail katydid takes momentary respite from its cage, crafted by entomologist-artist-musician Natalia Vandenberg (background). Insects provide inspiration along many different pathways.

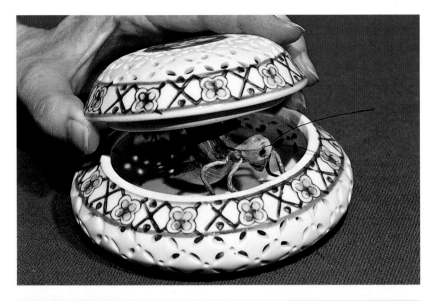

Top: This contemporary Chinese porcelain vessel is used, in part, to house singing crickets. Inside resides an artist's rendition of an Asian singing katydid made of wood and papier-mâché.

Bottom: This bottle gourd was grown in a mold and topped with a wood and perforated imitation tortoise-shell stopper. These gourd cages are each designed for particular species of singers. The one pictured, which would house a singing cricket, can be carried in the pocket as a form of personal enjoyment.

had been produced some twenty-three centuries earlier. Even the flea has been honored, in the "Song of a Flea" by Moszkowski. Finally, need I mention the ever-popular and inspirational song "La Cucaracha," devoted to the insect we most love to hate, the cockroach?

Arts and Crafts

Insects have appeared in the pictorial arts from Neolithic etchings on bone and rock to the elegant Tiffany stained-glass lamps of a cen-

Wildlife photography, an art form undertaken seriously by few, has limitless possibilities for subjects in the insect world. Here sits Carll Goodpasture, the contributor of all the photographs for this book, as he attempts yet another twisted angle from which to create his insect portraits. Photograph by Eric Grissell.

310

tury ago. Asian artists, in particular, for thousands of years have figured insects into their brush paintings. Greek jewelry, dating from 700 B.C., prominently displays lifelike, three-dimensional flies in gold and miniature chariots pulled by flies carved in stone. Greek coins honored the honey bee. In the Dark Ages, monks illuminated manuscripts with images of insects and plants. The Dutch and French schools of the seventeenth and eighteenth centuries produced realistic still-life paintings complete with insects. The artist and designer E. A. Seguy produced dozens of decorative illustrations of insects during the Art Nouveau and Art Deco periods. His work served to inspire the production of textiles, end papers, wallpaper, and stained glass. M. C. Escher, the artist noted for his intricate geometric compositions, used insects in a number of his works. Salvador Dali frequently placed insects such as ants and grasshoppers into his paintings.

Insects are used not only as an inspiration for art, sculpture, and decoration, their physical bodies and by-products may be used in the process of creating art as well. The iridescent blue wings of morpho butterflies are worked into pictures by artisans in South America, and in Africa bits and pieces of many different colorful butterfly wings are formed into lifelike compositions. Metallic green beetle wings are incorporated within sculptures in Thailand and earrings in South America, where live beetles also may be encrusted with stones and jewels and tethered to a pin on a woman's blouse. Entire insects naturally encased in amber have been used to create many kinds of jewelry, including broaches, necklaces, and earrings. Beeswax has been used over six millennia in the process of lost-wax casting for bronze works and is still used today in casting silver and gold jewelry as well as in the manufacture of hand-printed batik textiles.

History Lessons

When it comes to the role of insects in human history, we may stretch our credulity a bit to believe some of the past events that have been associated with them. Suppose, for a moment, it is true that the Chinese inventor of paper, Ts'ai Lun, created his process by watching the paper wasp chewing bits of tree bark, mixing it with saliva, and layering the mixture into the carton that surrounds her nestlings. Would this mean that civilization as we recognize it, which is so dependent on the written word, ultimately might be the result of mimicking insect behavior? Perhaps.

Certainly the biblical plagues of locusts were as real two millennia ago as they are in Africa today (although the swarms are somewhat tamed by chemicals and proper timing these days). Figures suggest that 2.6 square kilometers (1.0 square mile) of locusts can consume nearly 227,000 kilograms (250 tons) of food a day. Some swarms have been measured that cover more than 1000 square kilometers (400 square miles) and range for weeks over thousands of kilometers. These locusts competed with humans for food, but there is little way to know what effects they might have had on the development of civilization. One has to imagine, though, that such tremendous appetites must have slowed down progress substantially, especially in the Middle East and Africa, where the swarms have historically been most common.

One-third of the population of Europe died during the Middle Ages as a result of bubonic plagues (Black Death), which are caused by a bacterium carried from rats to humans by fleas. By the end of the 1800s, it is estimated that 6 million American Indians had died from the same cause. Books have been written on the history of insects and disease; these books itemize the deaths of millions of soldiers and describe how wars' outcomes have been altered as a result (for example, Hans Zinsser's *Rats, Lice, and History*). These simply horrendous losses include more than 350,000 deaths in the American Civil War caused by microbes associated with mosquitoes, flies, lice, and fleas. Two to three million Russians died during World War I from louse-borne typhus. One tangible effect of such insect-borne trauma concerns Napoleon Bonaparte, who set off to conquer Russia with almost 500,000 men. Within six months of the attempt, Napoleon withdrew from Russia with his 20,000 remaining men. The rest had died in large measure from disease, especially louse-borne typhus. Imagine what effect there might have been in the last century if Russia was a French-speaking country and Communists had never risen to power.

By now, you may be saying to yourself "What does this little lesson in mass carnage have to do with appreciating insects? It sounds as if we should destroy every last one." To which I would answer, "The good that insects have done for the advancement of humankind is likely to be even greater than the obviously negative, yet we rarely even consider the good." We are content to roundly condemn every last insect, with a few exceptions, without a trial and by guilt of association. This chapter, and indeed this book, has been written to give insects a fair chance to display their mettle and to demonstrate that they do a great many things that we never think about. True, insects are not elephants, panda bears, or spotted owls. They are not cuddly and most are not pretty; but, when it comes right down to it, insects have had a tremendous influence on the history of our lives and the natural scheme of things.

Getting Even: Biting Back

If you do not like insects for some reason or, even more radical, you want revenge, then I have some good news. Recently, at least three important books on insect control have been written. They are cookbooks. "Wait," you ask cautiously, "books for cooking insects? What do you do with a cooked insect?" Well, you may be sorry you asked. More than 1000 species of insects are routinely eaten throughout the world—200 different species in Mexico alone. Humans have eaten insects for eons. Insects such as locusts, ants, and termites are commonly eaten. It is one way to solve the problem of overabundance.

If the idea of eating cooked insects makes you feel a bit queasy, consider this notion: you eat a lot of them raw every day, so what's the harm in cooking them? According to data provided in May Berenbaum's *Bugs in the System*, a peanut butter and jelly sandwich may contain up to fifty-six insect parts (insect parts also include insect droppings, just in case you might be wondering). But don't worry, not a bit. You are protected by the law of the land. According to the U.S. Food and Drug Administration, 0.14 liter (0.50 cup) of raisins may contain no more than ten whole insects, or the equivalent parts, as well as thirty-five fruit fly eggs. Now doesn't that make you feel a whole lot better? Of course, raisins are not the only foods regulated by the FDA, there are lots of them. You need not worry about what insect parts are in whatever tasty tidbit you are contemplating. Just assume they are in everything you eat and forget the worry. In your lifetime, you've already eaten enough bugs to sink a small tugboat—maybe even two. In fact, eating insects might even be one possible way to save rain forests and all the animals that live in them (and maybe the human species, too). We could harvest insects in the forests and eat them instead of burning down trees to make rangeland to feed cattle. It's just a thought, of course.

Insect Ranching

Insect ranching, about as direct an input into insect culture as one could have, is becoming a financially profitable business. Overseas, butterflies are propagated in greenhouses without depleting the natural populations (some of which are already pretty far gone), and these are sold to collectors all over the world. A company in Japan converted some of its vegetable vending machines to sell live stag beetles. It seems the company combined a traditional summertime hobby of Japanese children, beetle collecting, with a by-product of mushroom farming, beetles, to make money from what had been considered a pest insect. In the United States, wranglers are producing stock for insect zoos, themselves on the rise in various large cities and some botanical gardens. Some entrepreneurs are rearing local butterflies to release at weddings instead of throwing the traditional rice. It's quite romantic, really, to leave a wedding under a flutter of butterfly clouds. Even Priscilla Huff's *101 Best Small Businesses for Women* suggests that ranching beneficial bugs for gardeners can be a profitable road to success.

Just the other day, I read that a Japanese gentleman paid $90,000 for an enormously large form of a nonendangered stag beetle. If this keeps up, insects might become more profitable to grow than gooseberries. It might be that, given a little time and some capitalistic thought, the most avid gardener might turn his garden into a profit-making, insect-growing ground instead of the literal money pit it normally is. The gardener could sell insects to restaurants, perhaps, Japanese gentleman with deep pockets, or other gardeners who need some good bugs.

Given the proper incentive, I have a feeling we all might learn to appreciate insects a little more than we think we do.

My local public garden, Brookside Gardens, Wheaton, Maryland, held a living butterfly exhibit in which children and adults with childlike tendencies interacted with hundreds of locally grown butterflies. As this monarch shows, the results were both positive and decorative.

14

The Realistic Gardener

Some fifty years ago, I planted my first nasturtium seeds in the backyard of our small San Francisco apartment house. The yard was divided into two sections by a tall picket fence: one section I could play in and plant seeds; the other I was forbidden to enter except with the sprightly escort of our landlord. (I once watched as he leapt over the fence in a single bound—too impatient, I suppose, to unlock the gate.) With his approval, I was allowed to explore the confines of his well-fenced and awesome vegetable garden. Next to our apartment, in a large house with a large garden, lived a neighbor who let me explore at will—let me collect and examine bugs and spiders; showed me the only working worm farm I've seen before or since. Yet another kindly old man, who lived up the street, showed me how a three-bin compost system worked; it's still the best one I've ever seen.

I have been interested in bugs almost as long as nasturtiums. As a student of insects, I have spent fifty years in pursuit of my buggy passions (nearly thirty in a professional capacity). During the late 1950s and early 1960s, I was a technician who helped professors and their graduate students test chemical insecticides on pears and apples, walnuts and grapes. I didn't really care for the job because it appeared that much of the time I was the object of the experiment. Spraying noxious chemicals all over creation (and me) did not seem, even to a budding entomologist, like an intelligent thing to do. And Rachel Carson, bless her soul, was about the only one who was not afraid to look us all in the eyes, even the entomologists, and tell us what idiots we could be. I eventually switched to the study of insect systematics, where I could investigate species, how they were named, how many kinds there were, their life histories, and how the species were related. It seemed like a relatively safe, nontoxic approach to things.

With purely subjective hindsight, it now seems as if I recollect seeing more different kinds of insect life in the backyard of an urban garden than I did in the pear and apple orchards of the farms I helped douse in pesticides. This is entirely logical, when you think about it. The object of those rustic farms—the rural bliss to which some of us occasionally aspire—was to create a bug-free zone. Toxic chemicals did a fine job of that. But, in the city gardens, all I can remember was crushing the brown garden snails (an invasive, introduced pest of which there was an endless supply), looking for live bugs to play with, seeing the earth replenished by compost, learning that worms helped the soil, and generally remaining in ignorant bliss that the world was much bigger and more toxic than our backyard.

In my earliest youth, I was extremely fortunate that gardeners introduced me to the world of natural processes. I did not know, at the time, that I lived in a vast urban sprawl, but neither did the worms, the nasturtiums, the bugs, or the spiders. As a considerably older gardener and entomologist, I now realize that these urban settings were fairly simple ones, devoid of much diversity, and largely unnatural. Yet they still demonstrated one basic premise of life: biological processes happen. They happen anywhere and everywhere, however, the extent to which they can take place in our gardens resides within us, the gardeners of the world. We can create anything from a totally organic garden (100 percent biological) to a totally bug- and weed-free one (100 percent chemically dependent). Alternatively, we can adopt any one of a million increments between these poles.

It is amid extremes that the realistic gardener must decide which

Within the natural world, there is an inexplicably fragile balance between its inhabitants. Here, a docile, adult flower fly plays its unwitting part in the continuation of water avens. As a larva, the fly was the killer of numerous aphids. Tomorrow it may be the prey of an assassin bug. And it all happens without the helpful hand of the gardener.

path the garden will take. The decisions are not easy, nor are they getting any easier. Since about the mid-1980s, we have seen that extremes in naturalistic gardening—especially the meadow lawn—have drawn the wrath of the pristine-lawn crowds, not to mention city council lawsuits against the naturalistic evildoers. Conversely, the sincere herbaceous borderist is made to feel like a diseased criminal by the native-plant police. And even the native-plant police can be made to feel like crazed poachers by the don't-harvest-the-endangered-seeds crowd. It seems that no matter what a gardener does, it is wrong in someone's mind.

Personally, I don't care how people want to garden, as long as they do it with a maximum awareness for balance and an appreciation for the environment in which we all live. Whether it is a formal garden or a vegetable garden, a cut-flower garden or a prize-flower garden, we can still have insects in our gardens if we don't panic at the sight of every strange creature we see flying through our domain. Grabbing the power sprayer as a first resort is like using a flamethrower to cook a hamburger: it gets the job done, but who wants to eat the results? Our first reaction to the appearance of strange and unknown bugs should not be chemicals, it should be contemplation.

I have heard many people justify their dependence on pesticides by saying that if their gardens had no insects, then there would be no problems. "The only good bug is a dead bug," they say. But this rather simple thinking reveals an absence of forethought. Just as we know that a simple garden, lacking in complexity, is more prone to the harmful effects of any one insect, the simple chemical solution to our problems are also prone to harmful effects. In the case of insects, we are liable to create a triple whammy with the use of simplistic thinking: first, we create chemically resistant superpests; second, we kill off the less resistant predators and parasitoids; and third, we spray ourselves and our neighbors in the face.

In general, we gardeners are a fine lot of people and, as I've remarked before in other places, if the world were composed entirely of us it would most probably be better off. At most there might be some competition to grow giant pumpkins or a champion dahlia, but it is unlikely that this rivalry would ever evolve into open warfare, as does almost everything else we think about. And yet there are several areas in which some of us gardeners—and I must admit to having been guilty here—sometimes revert back to the normal, primitive instincts of our friends who do not garden. In our gardens, we have an overbearing desire for order, an overwhelming obsession with perfection, and an oversimplified concept of biological facts. Taken together, this combination can stop naturalistic processes dead in their tracks.

To mix metaphors, we gardeners want our beans, not our ducks, in a row. We wish to view our vegetables in military precision without so much as a semicircle cut in a leaf; our lawns in verdant splendor without a dandelion; our immaculate roses with no underplants to distract our attention; our bulbs in tidy clumps, or sometimes, heaven forbid, straight rows; our annuals in annual beds; our perennials, ever-blooming of course, in resplendent borders—tall to the back, short to the front; our ponds crystal clear, with not an algal strand to be seen; our shrubs guarding the foundation of our castle; our tall conifers in forestlike cathedrals and our short ones in bowling-ball globes; our hardwoods symmetrical and protecting the property; our fallen leaves blown away; our organic debris in trash cans, replaced by shredded forests; our weeds chemically peeled away never to return. We want order, we want perfection, and we want it now.

Many of us attempt, in one way or another, to believe in some notion of a perfect, orderly landscape. Perhaps, as Evan Eisenberg suggests in *The Ecology of Eden,* we are looking for Arcadia, a paradise on Earth. If this is so, we sometimes inadvertently create a hell instead. The more we simplify our gardens in an attempt to retain the perfect order, the more we increase our need to do something to regain the natural, biological balance that it has lost. A simplistic loss of order is nowhere more obvious than when we try to tame exotic species to do our bidding, which is the basic tenet of a garden. It is with our smaller garden efforts in mind, perhaps, that we can envision the large-scale versions of what we sometimes do to the environment that surrounds us.

In humankind's quest for perfection, we have introduced invasive, land-consuming plants into our environment to do one thing, but have found that they are much better at doing something else. Kudzu is one major example. Introduced to prevent erosion, the kudzu vine simply smothers entire ecosystems. Not unlike ivy in our gardens, perhaps, or the chameleon plant so highly touted by growers several years back. Insects, too, are introduced to do good. The gypsy moth was introduced to create a silk industry in the eastern United States, but as we all know the moth ended up compromising millions of acres of trees as well as causing entire forests to be blanketed with noxious chemicals. We humans are rather good at moving plants and insects from one place to another, but we're not so good at guessing the outcome of doing so—either in our garden or in our environment. Which is peculiar, really, because we do have some theoretical understanding of how invasions work in nature. We know that an introduced species of plant or animal can become dominant and do what it does best: spread. These organisms win, in part at least, because they are growing without the herbivores that normally feed on them in their native environment.

The fact that we move plants and insects to new areas comes naturally to humans, we've done it since the beginning of time. Ever since our ancestors placed that first bunch of wild rice or wheat seeds into a pouch or a reed basket and moved to new planting grounds, we've been spreading stuff everywhere, both purposefully and accidentally. The result for the plant or insect being moved is entirely natural, they will simply play by the rules of nature no matter where they end up. If they are aggressive and successful, they will strangle the life out of every growing thing beneath their sprawl. In the long run it will be a total disaster for all but the adaptable survivors, but for them, it's a wonderful life, a life full of exuberance, unfettered growth, and unlimited possibilities. If a better and more aggressive player comes along, then the game of competition begins anew and the best competitor will win.

We gardeners have an innate power to change things. We spend much of our time trying to make our gardens the way we want them, or at least imagine we want them. We have found that if we work very hard and devote enough time, we can change any piece of undisturbed land into a pristine park, a botanical garden, an arboretum, a stately vista, a productive vegetable plot, or even a backyard paradise. People often equate great gardens with works of art—Manet's, Monet's, Constable's—but the realistic gardener knows that a garden is more nearly like a three-ring circus—balancing acts, knife throwers, and clowns in baggy pants. In the complex world in which we live, we often don't know how to approach the role of ringmaster without becoming an overbearing taskmaster. We can, on occasion, make our gardens into a work of art, but only for a moment: gardens are artifices of biology, not pieces of canvas. No matter how hard we work or how long we work at it, gardeners cannot change the way biological or natural principles work. In that respect, we either work with nature or we work against it.

Although much of this book has been based on explaining what insects are and where they fit into the natural world, it has also explored where the gardener fits in as well. The gardener, the garden, the insects, and the principles of nature are all inseparable, at least if we desire to be realistic gardeners. With a minimum of knowledge, we may be better prepared to adapt ourselves and our gardens to work within the principles of nature, rather than against them. It is in our battles over insects (and weeds, which follow the same biological principles as insects) that we sometimes lose sight of our own power to alter things, often for the worst.

In the first part of this book, I surveyed the basic layout of insect life found in our eminent domain, the garden. I attempted to do so from an enlightened perspective, one without prejudice or scorn, without ethical discussions or belligerent tirades as to who may live and who must die. I examined insects as creatures with unique lives, with places to go, with things to do. I proposed that insects are merely the normal beings in a world in which we humans are the neutral observer, a world that exists neither to purposefully challenge us nor to reward us in any preconceived way, a world that exists simply because it does.

In the second part of the book, I described insects in the context of their interactions with the garden, its plants, and, as importantly, each other. I viewed the garden from its occupants' position. Plants and insects have interacted for hundreds of millions of years. Why should we gardeners feel compelled to change this situation in an hour or an afternoon? What are insects doing that challenges us so? I explored these interactions, again with the eye of the impartial observer.

Finally, in this, the third part, I examined how the gardener fits into the almost incomprehensibly complex world of plants and insects. It is gardeners, after all, who imagine themselves to be masters of the garden (we have even bestowed upon ourselves the honorary title Master Gardener). If we learned nothing else from our explorations, it was that such a notion is pure self-conceit. Once we gardeners discover how insects interact with the garden and each other, we should be free to construct a garden that favors insect-plant interactions to the betterment of plant, insect, garden, and gardener. Armed with knowledge, hopefully we will be able to cast off the tyranny of our fear and loathing of insects and become less inclined to meddle in places where we have little or no success.

Ultimately, the aim of this book has been to share with gardeners the notion that if we choose to follow some simple principles, we will not have to meddle with insects much at all. The trick to getting started is, in a single word, *diversity*. The more plant and habitat diversity we have, the more naturalistic our gardens will become, and we will not have to think about planting a butterfly garden, a bat garden, or a deer garden. The diversity of insects will increase and that will increase the diversity of birds, bats, frogs, lizards, and so on. The more diverse all these organisms become, the less likely that any one will be a particular nuisance and the more likely some will find ways to control the others. The more natural controls we have, the less will be our need to have artificial ones. At some point, we will have so many plants to think about

Two bees or not two bees? That is really the question. Two long-tongued bees sit in a desert poppy, looking out at the world. We gardeners must determine how we see insects in our gardens and, more importantly, how we view nature within the framework of our gardens.

that no one plant will become sacred. Eventually, all these interactions will take care of themselves without any input from us. And then we will be free of the garden and free to garden.

A famous biologist and student of human ecology once said, "We can never do merely one thing." This has been dubbed Hardin's law, but Garrett Hardin modestly calls it the first law of human ecology. Its meaning is, of course, that all things are interconnected in some way. When we humans do one thing, in reality, we are doing many more things than we know about (or in some cases want to know about). In this book I hoped to bridge the gap between insects and gardener, to lighten our workloads, to enlighten our brainloads, and to suggest that our gardens are the proving ground between our personal lives and a larger world with which we are all interconnected.

Additional Reading

Alcock, John. 1999. *In a Desert Garden: Love and Death among the Insects.* 1997. Reprint, Tucson: University of Arizona Press.

Berenbaum, May R. 1995. *Bugs in the System: Insects and Their Impact on Human Affairs.* Reading, Massachusetts: Addison-Wesley.

Borror, Donald J., and Richard E. White. 1970. *A Field Guide to the Insects of America North of Mexico.* The Peterson Field Guide Series. Boston: Houghton Mifflin.

Borror, Donald J., Charles A. Triplehorn, and Norman F. Johnson. 1989. *An Introduction to the Study of Insects.* 6th ed. Philadelphia, Pennsylvania: Saunders College Publishing.

Bowers, Janice Emily. 1993. *A Full Life in a Small Place and Other Essays from a Desert Garden.* Tucson: University of Arizona Press.

Buchmann, Stephen L., and Gary Paul Nabhan. 1997. *The Forgotten Pollinators.* Washington, D.C.: Island Press.

Cann, David. 1998. Review of *The Self-Sustaining Garden* by Peter Thompson. *The New Plantsman* 5 (1): 64.

Čapek, Karel. 1984. *The Gardener's Year.* 1929. Reprint, Madison, Wisconsin: University of Wisconsin Press.

Carson, Rachel. 1994. *Silent Spring.* 1962. Reprint, Boston: Houghton Mifflin.

Conniff, Richard. 1996. *Spineless Wonders: Strange Tales from the Invertebrate World.* New York: Henry Holt and Company.

Cox, Jeff. 1991. *Landscape with Nature.* Emmaus, Pennsylvania: Rodale Press.

Creasy, Rosalind. 1982. *The Complete Book of Edible Landscaping.* San Francisco, California: Sierra Club Books.

DeBach, Paul, ed. 1970. *Biological Control of Insect Pests and Weeds.* 1964. Reprint, London: Chapman and Hall.

DeBach, Paul. 1974. *Biological Control by Natural Enemies.* London: Cambridge University Press.

DeBach, Paul, and David Rosen. 1991. *Biological Control by Natural Enemies.* 2d ed. Cambridge, England: Cambridge University Press.

DeFoliart, G. R. 1997. An overview of the role of edible insects in preserving biodiversity. *Ecology and Food Nutrition* 36: 109–132.

Eisenberg, Evan. 1998. *The Ecology of Eden.* New York: Alfred A. Knopf.

Ellis, Barbara W., ed. 1996. *The Organic Gardener's Handbook of Natural Insect and Disease Control.* Rev. ed. Emmaus, Pennsylvania: Rodale Press.

Emmel, Thomas C. 1997. *Butterfly Gardening.* New York: Friedman/ Fairfax.

Environmental Protection Agency. 1994. *Pesticide Industry Sales and Usage, 1992 and 1993: Market Estimates Report.* Cincinnati, Ohio: U.S. Environmental Protection Agency. (Summarized in Cooper, Cathy, ed. 1994. *Pesticide and Toxic Chemical News* 22 [33]: 8–10.)

Evans, Howard Ensign. 1964. *Wasp Farm.* London: George G. Harrap and Company.

Flint, Mary Louise, and Steve H. Dreistadt. 1999. *Natural Enemies Handbook: The Illustrated Guide to Biological Pest Control.* Berkeley: University of California Press.

Frost, S. W. 1959. *Insect Life and Insect Natural History.* 2d rev. ed. New York: Dover.

Gordon, David George. 1998. *The Eat-A-Bug Cookbook.* Berkeley, California: Ten Speed Press.

Gould, Stephen Jay. 1998. An evolutionary perspective on strengths, fallacies, and confusions in the concept of native plants. *Arnoldia* 58 (1): 3–10.

Graham, Frank. 1984. *The Dragon Hunters.* New York: E. P. Dutton.

Hardin, Garrett. 1993. *Living within Limits.* New York: Oxford University Press.

Hart, Rhonda Massingham. 1991. *Bugs, Slugs and Other Thugs: Controlling Garden Pests Organically.* Pownal, Vermont: Storey Books.

Hogue, Charles L. 1987. Cultural entomology. *Annual Review of Entomology* 32: 181–199.

Hölldobler, Bert, and Edward O. Wilson. 1990. *The Ants.* Cambridge, Massachusetts: Harvard University Press.

Hoyt, Erich, and Ted Schultz. 1999. *Insect Lives: Stories of Mystery and Romance from a Hidden World.* New York: John Wiley and Sons.

Hubbell, Susan. 1993. *Broadsides from the Other Orders.* Boston: Houghton Mifflin.

Hubbell, Susan. 1998. *A Book of Bees: And How to Keep Them.* 1988. Reprint, Boston: Houghton Mifflin.

Huff, Priscilla Y. 1996. *101 Best Small Businesses for Women: Everything You Need to Know to Get Started on the Road to Success.* Rocklin, California: Prima Publishing.

Lutz, Frank E. 1941. *A Lot of Insects: Entomology in a Suburban Garden.* New York: G. P. Putnam's Sons.

Marinelli, Janet. 1998. *Stalking the Wild Amaranth: Gardening in the Age of Extinction.* New York: Henry Holt.

Miller, Gary L. 1997. Historical natural history: insects and the Civil War. *American Entomologist* 43 (4): 227–245.

Mound, Laurence, and Stephen Brooks. 1995. *Pockets Insects.* New York: Dorling Kindersley.

Price, P. W. 1997. *Insect Ecology,* 3rd ed. New York: John Wiley and Sons.

Olkowski, William, Sheila Daar, and Helga Olkowski. 1994. *Common-sense Pest Control.* Newtown, Connecticut: Taunton Press.

Ordish, George. 1985. *The Living Garden: The 400-Year History of an English Garden.* Boston: Houghton Mifflin.

Osler, Mirabel. 1989. *A Gentle Plea for Chaos.* New York: Simon and Schuster.

Pleasant, Barbara. 1994. *The Gardener's Bug Book: Earth-Safe Insect Control.* Pownal, Vermont: Storey Books.

Ramos-Elorduy, Julieta. 1998. *Creepy Crawley Cuisine: The Gourmet Guide to Edible Insects.* Rochester, Vermont: Park Street.

Ryan, Lisa Gail. 1996. *Insect Musicians and Cricket Champions: A Cultural History of Singing Insects in China and Japan.* San Francisco, California: China Books and Periodicals.

Schultz, Warren, ed. 1995. *Natural Insect Control: The Ecological Gardener's Guide to Foiling Pests.* 21st-Century Gardening Series, no. 139. Brooklyn, New York: Brooklyn Botanic Garden.

Sear, Dexter. *Bugbios.* http://insects.org/index.html.

Shixiang, Wang. 1993. *The Charm of the Gourd.* Hong Kong: Next Publication.

Shorthouse, Joseph D., and Odette Rohfritsch. 1992. *Biology of Insect-Induced Galls.* New York: Oxford University Press.

Starcher, Allison Mia. 1995. *Good Bugs for Your Garden.* Chapel Hill, North Carolina: Algonquin Books of Chapel Hill.

Stein, Sara. 1993. *Noah's Garden.* Boston: Houghton Mifflin.

Swain, Roger B. 1994. *Groundwork: A Gardener's Ecology*. Boston: Houghton Mifflin.

Thompson, Peter. 1997. *The Self-Sustaining Garden*. London: B. T. Batsford.

University of Florida Book of Insect Records. http://www.rochsec.vic. edu.au/pages/Science/Insects/recbk.htm. 1997.

van den Bosch, Robert, and P. S. Messenger. 1973. *Biological Control*. New York: Intext Educational Publishers.

von Frisch, Karl. 1953. *The Dancing Bees*. 5th rev. ed. New York: Harcourt, Brace and Company.

Waldbauer, Gilbert. 1996. *Insects through the Seasons*. Cambridge, Massachusetts: Harvard University Press.

Waldbauer, Gilbert. 1998. *The Handy Bug Answer Book*. Detroit, Michigan: Visible Ink Press.

Wilson, Edward O. 1992. *The Diversity of Life*. New York: W. W. Norton and Company.

Winston, Mark L. 1997. *Nature Wars: People Versus Pests*. Cambridge, Massachusetts: Harvard University Press.

Xerces Society. 1998. *Butterfly Gardening*. Rev. ed. San Francisco, California: Sierra Club Books.

Zinsser, Hans. 1960. *Rats, Lice, and History*. 1934. Reprint, New York: Bantam Books.

Index

Entries marked in *italics* refer to photographs; those in **boldface** refer to the charts of Garden Arthropods (pages 67–70).